# From Speed to Cores: How Multi-Core Processors Changed Computing

**Nadal**

# Contents

# 1 Introduction

Hitting the power and heat walls has hindered the increase of clock speeds in single-core processors. To overcome these problems, the hardware industry has switched to multi-core processors and the success of multi-core processors has shrunk the market of single-core processors considerably. During the past decade, the degree of parallelism available in the hardware has grown quickly and decisively.

This change in the hardware industry has put pressure on the software community to develop parallel programs. Development of a sequential program for a single-core computer is not perplexing because the execution order of the program is obvious. But to take advantage of a multi-core architecture, programmers are required to develop programs that exhibit parallelism. Developing a parallel program is not simple, as a programmer has to take care of many aspects like data dependences, data races, and deadlocks. The problem is exacerbated if one's job is to parallelize a sequential program. A programmer should have deep knowledge of every code section of the program before starting the parallelization process. This process takes more time and energy if the sequential program being parallelized has been developed by someone else.

The problem of parallelizing a sequential program has urged researchers to invest in automatic parallelization tools. In recent years, the compiler community has responded to the need for automatic parallelization. Detection of parallelism in simple loops statically has been promising [4]. State-of-the-art compilers such as the Intel compiler [5] can automatically detect such parallelism. However, a wide range of sequential programs cannot profit from these strategies because such compilers may miss coarser-grained but sometimes more scalable parallelism. This is mainly due to the limitations of the static analysis used by these compilers. The static analysis is conservative in nature as it can not decide some dependences that occur at the runtime, like due to the complex array indexing.

To strengthen programmer productivity, over the years software engineers have collected a comprehensive list of sequential design patterns [3]. These patterns improve the quality, re-usability and maintainability of the software. Similarly, to simplify the development of parallel programs, many design patterns for parallel programming have been introduced [1]. These patterns provide reusable solutions for common problems faced during parallel programming and help avoid concurrency bugs like deadlocks and data races, which are very difficult to locate. These patterns are usually implemented via standard parallel programming interfaces such as OpenMP [6] or Intel TBB [7]. Jahr et al. [8] presented a systematic pattern-based approach to convert an existing sequential program into a parallel one. However, their method requires intimate familiarity with the code to be parallelized as well as comprehensive knowledge of parallel constructs and patterns.

This book presents a framework for the identification of design patterns for parallel applications in sequential programs. We use the dependence graph of a sequential program, output of our dependence profiler DiscoPop [2, 9, 10] (Chapter 3), as an input to our approach and try to identify different parallel patterns in it. Chapter 3 also gives an overview of our pattern identification framework. We employ three techniques: template matching to identify pipeline and doAll patterns

(Chapter 4); regression analysis to identify multi-loop pipelines, reduction and geometric decomposition patterns(Chapter 5); and breadth-first search to identify task parallelism (Chapter 6). Design patterns are introduced in detail in Chapter 2.

In the rest of this chapter, we first discuss the problems faced in parallel programming and why tools to identify parallelization opportunities are important. We present the types of parallelism available in programs before briefly introducing design patterns for parallel programming. We explain shortcomings of state-of-the-art parallelism discovery tools and how our approach tries to overcome them. In the end, we provide a brief overview of our approach and summarize the contributions of this book.

## 1.1 Era of Parallelism in Computers

The processing power of digital computers has been increasing continuously since their invention. The computer hardware industry has made many improvements in computer architecture. The number of transistors available on a microprocessor and its clock speed have increased significantly. Gordon Moore predicted in 1965 that the number of transistors per square inch on an integrated circuit would double every two years [11]. Since the 1980s, these new transistors helped in improving the clock speed of a microprocessor. Interestingly, the Dennard scaling law [12] states that as the transistors get smaller, their power consumption also decreases proportionally. Hence, the energy requirement for per square inch of the integrated circuit remain constant. However, by mid 2000s, the Dennard scaling law started fading away as the power could not be reduced comparatively to the reduction in the size of transistor [13]. The reason is the loss of static charge in transistors due to their very small sizes. Therefore, single-core microprocessors were hit by this power wall, where more power could not be provided to increase their clock speed. So improving the computing performance by just increasing the clock speed of a single-core processor was not possible.

To overcome the power wall faced by the single-core architecture, the microprocessor manufacturers came up with multi-core architectures. A multi-core processor has more than one independent processing core. All of these cores are capable of executing instructions simultaneously to each other. Moore's law still holds as the number of transistors on the processors are continuously increasing. However, the new transistors are now being used to add extra computing resources for parallel computing. A program that exhibits parallelism can take advantage of multi-core processors to increase its throughput.

A sequential program running on a multi-core processor cannot benefit from the parallel resources available. It still uses one core out of many for the execution. To take full advantage of parallel hardware, programmers need to implement parallelized programs. Developing a parallel program with simple paralellizable loops is straightforward but the complexity arises if the parallelism available in the program is not just confined to the loops but includes general programming constructs like functions. A programmer needs to understand the algorithm thoroughly before parallelizing the program to avoid concurrency bugs. This task is impossible without the help of tools if the structure of the program is very complex.

State-of-the-art compilers can transform simple loops into parallel ones automatically but they fail to identify the parallelism opportunities available in other constructs of a program. On the other hand, some tools, like Pluto [14], requires the programmers to provide hints in the input programs using annotations to automatically parallelize the annotated part of the program at compile time.

But again, the programmer needs to be careful about providing hints to the compiler or the resulting parallel version maybe not correct. Some approaches, for example Prospector [15], try to identify only a single type of parallelism like pipeline or complex doAll loops but they miss other types of parallelism opportunities.

Software engineers have introduced a number of design patterns for parallel programs. These patterns provide reusable and bug-free solutions for common parallelization scenarios encountered by programmers. However, applying these patterns to parallelize a program is not a straightforward task. Identifying an appropriate pattern for a code segment of the program to achieve best speedup is very difficult. A programmer has to master the patterns and also acquire deep knowledge of the program.

This book presents a framework consisting of various approaches that identify different design patterns for parallel programs in a sequential program. Our approach not only identifies a design pattern but also specifies the division of code blocks according to its structure. Based on this information, the programmer can then easily parallelize the code by transforming recommended code blocks as suggested by the pattern.

## 1.2  Types of Parallelism

There are many types of parallel hardware available on the market. Flynn classified computer architectures into four classes based on the number of instruction and data streams they can handle [16]. This classification by Flynn is also known as Flynn's taxonomy. These four classes are:

- *Single Instruction stream, Single Data stream (SISD)* computers are the typical sequential computers. These computers have only one single-core processor that can work on a single instruction at a time, operating on a single data stream.

- *Single Instruction stream, Multiple Data stream (SIMD)* computers have multiple processors. All the processors execute the same program over different streams of data. A modern day processor with vector instructions is an example of such a system. These systems are very efficient for executing the same operation on different elements of an array in parallel.

- *Multiple Instruction stream, Multiple Data stream (MIMD)* computers are the modern multi-core systems. These are general-purpose systems with each processing core having its own instruction stream that it executes on its own separate data stream. All cores are independent of each other and work in parallel.

- *Multiple Instruction stream, Single Data stream (MISD)* computers are very uncommon. They execute multiple instructions on a single data stream. Systolic arrays [17] are a good example of such a system. Their architecture contains a network of processors, that apply different operations on the input data stream and merge the results in the end.

The parallelism in software is generally categorized into two main types. *Data parallelism* can be achieved by distributing the data between different processing cores. McCool et al. [18] define data parallelism as any kind of parallelism that grows with the growth of the data set. Data can be split into different sizes of chunks according to the semantics of the program. Some programs process the input data in parallel by simply breaking it down into discrete blocks like different parts of an array. Other programs may traverse the data recursively like processing different branches of a tree in parallel. In data parallelism, the sequence of instructions performed on all chunks may be same (SIMD) or they may differ according to the data in a chunk (MIMD). Data parallelism is termed as the most scalable form of parallelism [18] as the bigger data leads to higher number of chunks, hence creating more parallelism opportunities.

The second type of parallelism is known as *functional parallelism*. This parallelism is achieved by executing the different functions of a program in parallel to each other. Functional parallelism can only be implemented if the program contains some functions or code sections that are not dependent on each other. To get the best possible speedup, all functions should have a sufficient and almost equal amount of load. This type of parallelism is often not as scalable as data parallelism, because it depends on the number of individual functions. However, functional parallelism can help achieve better performance when used in conjunction with data parallelism as an extra parallelization opportunity.

## 1.3 Design Patterns for Parallel Programs

Design patterns were first introduced by Alexander et al. in 1977 in the field of architectural design [19]. Architects have to solve the design problems they face when planning to build structures, such as, offices, homes, cities etc. The main challenge is to ensure better accessibility to all facilities for the end users. Most of the time, the problems in designing new projects are similar to the ones that architects have encountered before in the other projects. A design pattern provides the core of the solution for a specific recurring architectural design problem.

Design patterns provide solutions in the form of reusable templates. These solutions can be applied multiple times, wherever the problems addressed by their respective design patterns occur. The exact details of a solution in a project depends on many different specific conditions present on the ground. Due to this, the final outlook of the solution may be different from the one implemented in other projects for the same problem. But the basic structure of the solution remains the same. Therefore, design patterns not only save architects from reinventing the wheel but also ensure that the basic issues of the respective problem are addressed properly. For example, one pattern named *City Country Fingers*, provided by Alexander, deals with the problem of designing a city in a way so that people get access to market places as well as farmlands within walking distances. The solution of this pattern proposes a design with multiple suburbs in the shape of fingers extending into farmlands. However, the actual realization of the proposed solution heavily depends on the geographical constraints of the city and will result into a different architecture than another city built using the same pattern.

Software engineers picked up the idea of design patterns from the work of Alexander et al. and applied it to the software design process [20]. The idea became particularly famous for designing object-oriented applications. Gamma et al., also known as the *Gang of Four*, published a list of reusable software design patterns that helped make the process of object-oriented software design easier [3]. They defined design patterns on an abstract level so that these patterns can be used in software of any type of domain and for any programming language. Gamma et al. described software design patterns with four essential elements: pattern name, problem description, solution, and consequences. The first three components are quite obvious with their names and the fourth one states the results and trade-offs of using the respective design pattern.

Software design patterns have many advantages in the software development cycle. Design patterns are usually very well documented and are understood by most of the software engineers. The usage of these patterns makes the software design of an application more understandable. Each pattern usually has more than one tried and tested solution available. An implicit benefit of using a well tested solution is the reduced risk of failure and bugs in the project.

Based on the remarkable success of software design patterns for object-oriented applications, software engineers introduced design patterns for parallel programming. For the sake of simplicity, we call them *parallel design patterns*. Many researchers proposed different ideas of using design patterns in the field of parallel programming [21, 22, 23]. However, the work done by Mattson et al. [1] was one of the first in this direction. They defined four categories of parallel patterns and classified many patterns according to these categories. For example, the pipeline and recursive data patterns in the algorithm structure design space. We will discuss these categories and their parallel patterns in detail in Chapter 2.

Design patterns for parallel programming carry all the advantages similar to the object-oriented software patterns. They help in making the parallelization process of a program easier. The solutions provided by parallel patterns simplify the implementation of a parallel application. Debugging a parallel application is a very hard task due to non-deterministic execution of parallel code. Unstructured parallelization can lead to many concurrency bugs like deadlocks and data races. Usage of parallel patterns to parallelize an application implicitly avoids these bugs as the solutions available in these patterns are designed by experts and have been rigorously tested in different scenarios.

## 1.4 Parallelism Discovery

Automatic discovery of parallelism has been the target of many researchers for the last decade. The idea is to take the burden of parallelization off a programmer's shoulders and develop a tool that produces highly efficient parallel applications from sequential code without any external hints or help. However, discovering parallelism in sequential programs is not a straight forward task. A programmer writing a sequential program implicitly provides a strict order of execution for program statements. A statement may use the results of any number of previous statements in the program, hence creating dependences among the statements.

The main hurdle in the parallelization process is to overcome dependences. Dependence resolution enables the execution of some statements reordered, so that they can run in parallel with other statements. This reordering should ensure the correctness of the whole program. A programmer having deep knowledge of a program may resolve some dependences by changing the algorithm structure or using some other constructs to achieve parallelism. Performing such type of dependence resolution can be very difficult for an automatic parallelization tool without any external hints because the tool may not realize which statements can be reordered without affecting the semantics of the program.

There are two type of dependences: data and control dependences. Data dependences are the most important elements to be taken care of when parallelizing a program. A correct parallel program needs to ensure that the production and consumption of data is in correct order, otherwise it can produce wrong results. Kennedy and Allen [24] state that a data dependence exists between two program statements $S_1$ and $S_2$ if and only if:

- $S_1$ and $S_2$ access the same memory location and one of them writes into it.

- There is a feasible runtime execution path from $S_1$ to $S_2$.

Based on this definition, three types of data dependences can be deduced:

- **Flow dependence:** If $S_1$ writes to a memory location before it is read by $S_2$. A flow dependence is also know as *read-after-write* (RAW) dependence.

- **Anti dependence:** If $S_1$ reads from a memory location before it is written by $S_2$. An anti dependences is also called *write-after-read* (WAR) dependence.

- **Output dependence:** Both $S_1$ and $S_2$ write into the same memory location. An output dependence is also called *write-after-write* (WAW) dependence.

Resolution of WAR and WAW dependences for parallelizing a program is usually straightforward. These dependences can be resolved by using a separate memory location for writing the new value. This is accomplished by using a separate variable name or making that variable private to the threads running the parallel program. The resolution of RAW dependences is a difficult task and most of the times they cannot be resolved. Two or more program regions can truly run in parallel if there is no RAW dependence between them. If such a dependence exists, then the communication needs to be synchronized between the parallel regions. This creates an overhead and affects parallel efficiency.

Similar to data dependences, a control dependence exists if the execution of a program statement is dependent on the outcome of another statement. According to Kennedy and Allen, a statement $S_2$ is control dependent on a statement $S_1$ if:

- There exists a non-trivial path $P$ from $S_1$ to $S_2$ such that every statement $S_i \neq S_1$ is post-dominated by $S_2$.

- $S_1$ is not post-dominated by $S_2$.

A node $x$ in a flow graph is post-dominated by a node $y$ if all the paths starting form $x$ to an *exit / destination* node necessarily pass through $y$.

The example code in Listing 1.1 can explain a control dependence easily. In this code segment, there exists a control flow path from statement A to statements B and C. Also A is not post dominated by B and C but by statement D. Hence, the statements B and C are control dependent on the statement A.

Listing 1.1: An example code to explain control dependences.

```
1 A:  If (a<0){
2 B:      x--;
3 C:      y--;
4 }
5 D:  x++;y++;
```

All the programming language constructs, that are used for decision making in a program, for example if, switch and loops etc., create control dependences. Kennedy and Allen discussed two techniques to deal with control dependences when parallelizing a program. The first one is called *if conversion*, where if conditions are modified into data dependences. This technique leads to vectorization of the code in some cases. The second technique is to extend the dependence graph of a program to include control dependences along with data dependences. This technique avoids any bugs in parallelization where the dependence graph is used as the basis.

A more modern technique to handle control dependences is speculative parallelism. In this technique, the control dependent statements are run in parallel with the conditional statement, speculatively. The original values of the variables changed in dependent statements are also maintained, separately. The end values of these variables depend on the result of the conditional statement. Either new values or the original values are propagated. For example , in the above code, statement A can run in parallel with statements B and C on separate cores. However, based on the result of A, the correct values for variables x and y will be used in statement D. Speculative parallelism can be achieved by using either speculation supporting software or hardware.

Many automatic parallelization tools have been developed that identify parallelization opportunities by analyzing data and control dependences of a program. These tools can mainly be divided into two categories based on how they obtain these dependences, namely: static and dynamic methods. We discuss each of these categories in detail in the following subsections.

### 1.4.1 Static Methods

In static methods, the dependences are inferred by just analyzing a program at compile time. This approach can obtain the dependences for all the code sections of the analyzed program. However, there are some drawbacks in static methods. Due to the lack of runtime information, deducing some dependences, like those caused by pointer aliasing or complex array indexing, may not be possible. Also, if the program under analysis is too big, then exploring all control flow paths can face a problem called *branch explosion*. Below, we discuss some tools that identify parallelism in sequential programs based on static analysis.

#### Polly

Polly [4] is an automatic parallelism detection tool based on LLVM [25, 26]. It uses a mathematical model called the *polyhedral model* [27]. The polyhedral model translates a program into a mathematical representation that supports the optimization and parallelization of loops and loop nests. Polly applies the polyhedral model on the low level intermediate representation (IR) of a program in LLVM.

The Polly front-end detects program regions where the polyhedral model can be applied. These regions are called *Static Control Parts* (SCoPs). The detected SCoPs are then translated into the polyhedral representation and dependences are inferred statically using ISL flow analysis [28, 29]. Afterwards, different loop optimizations and parallelizations are performed. Polly also supports the integration of other tools that can further optimize or parallelize the detected SCoPs. In the end, the optimized polyhedral representations are converted back into LLVM IR. Polly supports the OpenMP-based parallelization of loops. The converted IR is compiled into optimized and parallelized code.

Grosser et al. [4] tested Polly on the PollyBench 2.0 benchmark suite [30], containing thirty benchmarks from different fields like computation kernels used in linear algebra, data mining, stencil and image processing. Polly generated OpenMP based parallel code automatically for the detected parallelizable SCoPs in these benchmarks and was able to achieve an average speedup of $12\times$ on 24 threads. Polly achieved $80x$ and more than $100x$ speedups in two cases by combining parallelism with other optimizations like vectorization and tiling.

For simplicity reasons, a SCoP in Polly can contain only `for` loops and `if` conditions. The loop induction variable should be incremented by a constant value and its values should be bounded by affine parameters, which should not change during the runtime of the loop. Such definition of a SCoP is very strict. Many loops and regions in a general purpose programs may not qualify as a SCoP for Polly. This can result in missing opportunities for some potential parallelism available in those missed out regions.

### AutoPar

AutoPar [31] is an automatic parallelism discovery tool and part of the ROSE compiler project [32, 33]. ROSE is a source-to-source compiler developed by Lawrence Livermore National Laboratory. It supports transformations and analyses of C/C++, Fortran, UPC, Java, Python and PHP based applications. AutoPar works on high-level abstractions, such as vector, set and linked list etc. It supports both the standard template library (STL) of C++ and user-defined data structures.

AutoPar is a semantics-aware parallelizer. These semantics are provided by the programmer as an input to the compiler. This information tells the compiler whether the elements of an abstraction are alias-free and non-overlapping. The programmer also provides the side effects of each member function of the abstraction, stating if it only reads the elements or also modifies them. All this semantic information is used by AutoPar to identify parallelizable loops in a program that use these abstractions. Here are the steps taken by AutoPar to parallelize an input program [31]:

1. Read the user-defined semantic information.

2. Try to transform the source code by using optional optimizations, for example, convert list traversal into loop iterations. Also, normalize loops using iterators.

3. Find loops doing array computations suitable for OpenMP `for` and `task` constructs.

4. Skip loops that call functions with side effects or where function semantics are not provided.

5. Perform liveness and dependence analyses.

6. Classify the variables according to the OpenMP scopes and eliminate dependences resolved because of the variable scoping.

7. Insert OpenMP constructs once all inter-iteration dependences are resolved.

The dependence analysis of AutoPar categorizes variables referred in each statement into a read set and a write set. The references returned by function calls are checked for semantic equivalence with an array-like accesses by using already known abstractions from the abstract syntax tree of the ROSE compiler. All the variables affected by the modifying functions are collected, too. The loop induction variables are eliminated using a Gaussian elimination algorithm. At the end, a dependence relationship is generated between the two references of a variable in different statements if one of the references is a write operation.

The evaluation of AutoPar on different general purpose applications shows an almost linear speedup in most of the cases [31]. If provided with an OpenMP parallelized program, AutoPar is able to check the correctness of the program. However, this correctness analysis of AutoPar is heavily dependent on the provided semantic information of abstractions by users. This can be very inconvenient for a programmer, especially if the program under analysis is too big or very complex or is developed by someone else.

## Intel® Compiler

The Intel compilers for C++ and Fortran [5, 34] can auto-parallelize a loop if it is deemed safe to be parallelized. These compilers analyze the loops in the program being compiled and look for loops and loop nests with very simple structures. Basic requirements for a loop to be auto-parallelized by the Intel compiler are the same as the one for OpenMP loops: the number of iterations should be known; no breaks inside the loop body; and no inter-iteration dependences.

Intel compilers do not auto-parallelize the loops if they have any aliased pointers or complex array indexes. Additionally, any calls to external functions inside a loop also disqualifies that loop for auto-parallelization, as these functions may have side effects and can cause inter-iteration dependences. Intel compilers are by default conservative and do not parallelize a loop if they find any of the aforementioned reasons in the loop. However, Intel compilers have built a mechanism to auto-parallelize such loops with the help of programmers. A programmer can explicitly tell the compiler that the aliased pointer, the array index, or the called function are safe to be parallelized.

Intel compilers have a sophisticated reporting mechanism that provides the information about the loops that have been parallelized. They also report the loops that cannot be parallelized and provide the reason that blocked the auto-parallelization process. The programmer may change the loop structure to overcome this hindrance, so that the compiler can parallelize the loop.

## The Paralax Infrastructure

Vandierendonck et al. [35] introduced a framework that transforms a sequential program into a parallel one semi-automatically. The main focus of the framework are irregular pointer-intensive C programs with data structures. It has the following four components:

1. The Paralax compiler that auto-parallelizes coarse-grained loops operating on whole data structures. Such loops have large iteration spaces and alias analysis has more probability to succeed by grouping memory references per data structure.

2. The Light Weight Programming Model (LWPM) that introduces annotations in the code. These annotations are helpful in providing semantic information about variables, function arguments, and functions to the compiler.

3. Methods that convert the annotations into code fragments. This helps check the correctness of the annotations provided.

4. A tool that proposes locations to insert annotations in the sequential code. This tool uses dynamic dependence analysis during the profile run and compares them with the statically determined dependences. The tool analyses the differences between the two and suggests insertion of annotations in the program code. This helps in enhancing the parallelism opportunities in the sequential program.

For static dependence analysis, Paralax uses a mixture of static single assignment (SSA) for data structures and use/def chains. For parallelism detection, nodes of the program dependence graph of a loop are merged into strongly connected components (SCCs) that later on become stages of a pipeline. The dependences between the SCCs are relaxed as much as possible using information provided by the programmer through program annotations. Paralax also supports OpenMP sections-like parallelism.

The evaluation of the Paralax infrastructure shows speedups in the range of $1.8\times$ to $7\times$ using eight cores. The main reason for such limited speedups is that Paralax parallelizes the loops with pipelines. A pipeline with big sequential stages does not scale well.

**Parcae**

Parcae [36, 37] is a compiler and runtime software system that delivers performance portability for general-purpose and array-based programs. The Parcae compiler, called Nona, can detect multiple types of parallelism in loop nests of the sequential program. It can identify doAll, doAny, and pipeline parallelization opportunities. It uses the same dependence analysis as Paralax and supports dependence relaxation based on annotations provided by the programmer.

Nona generates multiple versions of parallelized code for each identified opportunity. Each version exhibits a distinct type of thread-level parallelism. The sequential version of the code is also compiled and kept in the executable of the program. For monitoring performance at runtime, the compiler instruments the program. This instrumentation is later used by the runtime system. The compiler also inserts code to pause, reconfigure, and resume the execution of the parallelized loop nest.

The runtime system of Parcae consists of the Decima monitor and the Morta executor. Decima uses the performance data produced by the instrumented program and also keeps track of system changes like new programs being launched or some hardware going off-line. Based on the analysis of all the data available to Decima, it can instruct Morta to pause the currently executing parallel version and switch to another suitable version produced by Nona or change the system configuration. The goal of the runtime is to maximize the overall system performance. If none of the parallelized versions or configurations result in better performance, the runtime system can always revert to the sequential version of the loop.

**Others**

Parallelism discovery using static methods is an active area of research. The above mentioned tools are some examples. Besides these tools, Li et al. [38] parallelize a sequential program by converging its program dependence graph into strongly connected components. They further coarsen the components using typed fusion to further reduce the synchronization overheads. Their dependence analysis lacks support for vectors and reports dependences only on scalar values.

Like Polly, Pluto [14, 39] is also an automatic parallelization source-to-source translator. It uses the polyhedral model to identify parallelism in loop nests. A programmer needs to annotate the potentially parallelizable candidate loops of a sequential input program. Taskminer [40] employs static analysis to automatically annotate sequential programs with OpenMP task pragmas. The conservative nature of the static analysis may cause some parallelization opportunities to be missed by them. Yucca [41], previously known as S2P, generates OpenMP and pthreads-based parallel programs automatically for a sequential input C program. It uses rigorous dependence analysis and detects the sections of code that can run in parallel. Yucca reports parallelism only between tasks formed from constructs contained in the main function of the C program.

### 1.4.2 Dynamic Methods

As discussed in Section 1.4.1, static methods have some inherent problems due to their lack of actual runtime information. Dynamic methods overcome the drawbacks of static methods by instrumenting the program and gathering the dependence information at runtime, based on the actual access to memory locations. But dynamic methods also have some disadvantages. The outcome of the analysis depends on the input provided to the instrumented program. It may happen that some code segments of the program may never run due to the nature of the input data. Also instrumenting a program increases huge time and memory overhead on the program execution.

Many researchers have tried different approaches to identify parallelism using dynamic methods. We discuss some of the state-of-the-art approaches in the following.

#### Prospector

Prospector [15] uses dynamic dependence analysis to discover parallelism in loops of the sequential input program. The input to Prospector is the source code or the binary files of the sequential program. Prospector performs control-flow analysis to detect loops at compile time. Instrumentation of memory-access instructions in loops is also done in order to perform dependence analysis at runtime.

Prospector only analyzes loops for dependences instead of the whole program. The dependence analysis is done at runtime. To reduce the memory overhead, Prospector uses a stride-based compression technique. Strides are detected in the memory-access patterns of loops and used for dependence analysis instead of single memory locations. Prospector is able to detect the source and sink of a dependence, its frequency, and its distance. It also profiles complex control flows including the dynamic extent of a loop. To achieve time efficiency, Prospector splits the memory space used by a program among multiple threads. Each thread is responsible for dependence analysis of its own memory space.

Based on the results of dependence analysis, Prospector can identify all three types of inter-iteration dependences (RAW, WAR, and WAW) in a loop. The output of Prospector shows potentially parallelizable loops that have no or very few inter-iteration dependences. A programmer can use this information to parallelize the program easily.

Prospector profiles only loops in the input program and ignores all the code sections outside loops. Prospector does this to avoid big memory and time overheads. Most of the execution time in a program is consumed by loops and they are good candidates for parallelism. But, still many applications may have big parallelization potential outside loop constructs, such as functional parallelism. In these cases, Prospector shall miss the optimal parallelism strategy that can be employed to parallelize a program.

#### Tareador

Tareador [42, 43] supports parallelization using task decomposition. The input to Tareador is an annotated sequential program, with annotations marking the start and end points of potential tasks in the program. Tareador uses Valgrind [44] for profiling the suggested tasks and detects data dependences between them. The slowdown due to profiling is reported to be between $200\times$ to

1000×. The output of Tareador is a dependence graph of the defined tasks. Detected tasks without any dependences between them can be run in parallel to each other.

Tareador uses an automatic iterative process to decompose a program into tasks. In the first iteration, the whole program (i.e. `main` function) is taken as a task and then further decomposed into smaller tasks based on its child constructs (loops and function calls). At each iteration, the estimated speedup is calculated and if it is better than a specified value, the process ends. Otherwise, one of the bottleneck tasks is selected heuristically and further decomposed into smaller tasks. A potential bottleneck is a task with longer execution time, too many dependences, or low instance concurrency (i.e., low number of dynamic instances executing in parallel).

Tareador can help parallelize applications exhibiting functional parallelism. The results show that Tareador achieves around $3\times$ speedup on 4 cores for many small applications. However, Tareador may perform very badly in identifying parallel tasks where all the potential tasks depend heavily on each other.

**Parceive**

Wilhelm et al. [45] have developed a parallelism identification and optimization tool called Parceive. It uses Pin [46] to instrument program binaries for profiling both sequential and parallelized applications. Parceive profiles the following data in a program:

1. Function calls and returns to keep track of the execution time of each function. Parceive also gathers the thread identifier in the case of multi-threaded applications.

2. Memory accesses along with the type of access and variable names for dependence analysis.

3. Memory allocation and release function calls for liveness analysis.

4. Calls to APIs creating and destroying new threads to keep track of thread operations and synchronization.

All of the profiling data is recorded in a relational database. To reduce the instrumentation overhead, Parceive provides an interface to the programmer to exclude any part of the program from instrumentation.

The database created during the profiling part becomes the base for many analyses, such as data dependence, race detection, and parallelism identification. All analyses can be done using SQL queries on the database.

A visualizer component visualizes all the useful information retrieved by SQL queries in an interactive way like function execution time, data dependences between functions, number of accesses to a variable etc. There is no automatic parallelism identification in Parceive, but the interactive visualizer helps programmers to identify parallelism opportunities in the program. The case study presented by Wilhelm et al. shows the successful identification of a pipeline in a test benchmark.

Using a relational database to store profiling data is an interesting approach. Many types of analyses can be done by querying the database. However, this approach has a very big time and space overhead. The results showed that Parceive produced 502 MB of data for profiling BZip2 [47], used to compress a 100 KB text file. Profiling slowed down the execution by a factor of 812.

### Intel® Advisor XE

Intel developed a tool called Advisor XE [48] to help programmers parallelize their programs. The supported programming languages are C, C++, C#, and Fortran and it works both on Windows and Linux operating systems. The workflow of Intel Advisor XE consists of the following four stages:

1. Survey target: In this stage, the sequential input program is run and the tool detects hotspots in it. This is done by counting FLOPS and the execution frequency of each code section. This helps a programmer to focus on these hotspots to achieve the highest degree of parallelism. In this stage, the tool also analyzes the code for vectorization opportunities and provides extra hints to the programmer about why a specific code section could not be vectorized. The programmer can use these hints to resolve these problems and enable automatic vectorization for the respective code sections.

2. Annotate sources: Once hotspots have been detected and picked as potential candidates for parallelism, the programmer should annotate these hotspots. The tool provides suggestions for such annotations and help the programmer to sketch his parallelism strategy for each hotspot. These annotations do not make the input program parallel. They are used by the tool to analyze the program during its next run.

3. Analyze performance and scalability: The annotated code is run again by the tool and emulates the parallelism for each annotated code region. The results of this stage show scalability, potential load imbalance, lock contention, runtime overhead and expected speedup for each annotated hotspot. The programmer can select between different design patterns and models of parallelism and compare the benefits of each. This helps programmers to see the benefits of each parallelism opportunity before investing into actual parallelization of the code.

4. Check correctness: This stage looks for potential concurrency bugs like data races and deadlocks in the annotated code sections. For this, Advisor profiles the data dependences of the input program. The idea is to find and fix these bugs even before the actual parallelization of the program occurs. At this stage, the programmer can see all the places that may result in a parallelization bug and resolve them accordingly by using different synchronization techniques.

Once all four stages have been completed and many alternative parallelization strategies have been compared against each other, the resultant annotated program can easily be transformed into a parallel one. Any type of parallelization model like OpenMP can be used for this purpose. The annotations in the program tell the programmer how to parallelize a code section and where to add synchronization constructs to avoid concurrency bugs.

Intel Advisor XE is a good tool for making the parallelization process easier for programmers. But, the hard task of identifying the parallelization strategy is left to the programmer. Knowing the impact of parallelization decisions even before implementing them is a very helpful feature provided by Intel Advisor XE.

### Wang et al.

Wang et al. [49] developed an approach for the identification of parallelism in loops of a program using both static and dynamic methods. Firstly, they extract dependences of the input program using static analysis. Then they instrument the program to profile the memory accesses of the program. The data generated by running this profiled program is used to detect both data and control dependences dynamically. The dependences detected by both analysis are merged and mapped onto a program graph called *control and data flow graph* (CDFG). If a *may* dependence is not realized in the dynamic analysis, it is reported to the programmer for further verification. The CDFG of the program is the basis for parallelism detection.

Wang et al. detect doAll loops by investigating the absence of inter-iteration dependences in them. They produce a parallel version of the input program using OpenMP constructs. Their approach classifies a variable as private to a loop if a write to this variable happens before it is read inside the loop. They also detect reduction opportunities on scalar variables using an algorithm presented by Pottenger [50].

A machine learning approach is used to establish whether a loop is profitable to be parallelized. It also proposes a suitable scheduling clause for each parallelizable loop. The machine learning model is based on four features: IR instruction count; IR load/store count; IR branch count; and loop iteration count. The results reported by Wang et al. show significant improvements in comparison to the auto-parallelization of the Intel compiler. They reported an average speedup of $3.45\times$ on a Xeon processor over the auto-parallelization achieved bt the Intel compiler across different benchmarks.

### Others

Other than the methods discussed above,many parallelism detection techniques based on dynamic analysis have been proposed. Garcia et al. [51] developed a tool called Kremlin. Their approach to discover parallelism uses critical path analysis [52]. Kremlin uses hierarchical critical-path analysis for nested regions and quantifies the parallelism for each region. It can suggest parallelism opportunities in the sequential input program that are easily parallelizable using the programming models, like OpenMP and Cilk++.

Parwiz [53] profiles a program and stores the profiled data in an execution tree. Every access to a memory location is stored in the execution tree and data dependences are extracted from it. Parallelization opportunities are discovered by analyzing the tree and data dependences between the nodes of the tree. Parwiz also optimizes the profiling overhead by grouping contiguous memory accesses into a single block and using this block for reference instead of individual memory locations. Another tool called Alchemist [54] gives parallelization recommendations for the input program. It calculates the distances between the references to the same memory locations in a program region and the continuation of that region. Alchemist decides to parallelize a region based on these distances. It does not follow any specific programming model and the parallelization strategy decision is left for the users to make.

## 1.5 Preliminary Work

This section introduces the parallelism discovery framework DiscoPoP (Discovery of Potential Parallelism) [2, 9, 10]. The data gathered by DiscoPoP is used as he basis for this book. DiscoPoP is based on the LLVM compiler [26]. It uses the LLVM Pass framework [55] to analyze the input code and instrument it. DiscoPoP uses a hybrid analysis technique (i.e., both static and dynamic) to extract data and control dependences as well as control region information of a program.

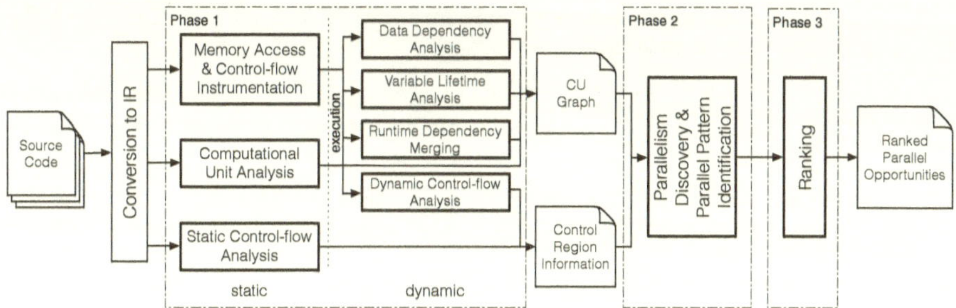

Figure 1.1: Parallelism discovery workflow.

This data is then further analyzed for the discovery of parallelization opportunities.

Figure 1.1 shows the workflow of DiscoPoP. There are three main phases of DiscoPoP: The first phase deals with dependence analysis. The memory accesses of the input program are instrumented and control flow information is extracted statically during the first phase. The instrumented program is executed to obtains data dependences. The second phase comprises parallelism discovery and pattern identification. This is done by analyzing the extracted dependences using different techniques including pattern identification. It also uses the information about the input program gathered in the first phase. In the end, the third phase sorts the parallelism opportunities discovered in the second phase according to different metrics like coverage, speedup, and load imbalance. We discuss each phase further in detail.

### 1.5.1 Phase 1: Dependence and Computational-Unit Analysis

Both static and dynamic analyses of the input program are done in the first phase of DiscoPoP. The input to this phase is an input program converted into the intermediate representation (IR) of LLVM [25, 26]. We use clang [56], a C language family frontend for LLVM. The static part consists of:

1. the instrumentation of the input program to profile the memory accesses and the control-flow. This instrumentation inserts library function calls before each memory access, function call, and loop. These library functions record their relevant events, which are used for dynamic analysis.

2. a static control-flow analysis that determines the boundaries of control regions like loops, if-else and switch blocks etc.

3. a computational-unit analysis that identifies *computational units* (CUs) in the input code. A CU is a block of code that follows the read-compute-write pattern [57] and has little parallelism inside it. Further details about CUs are discussed in Chapter 3.

The dynamic part of the phase consists of:

1. data dependence analysis of the program. It uses a signature based algorithm to detects data dependences in the program.

2. variable lifetime analysis to discard the variables not available in a specific code region. This improves the data-dependence accuracy.

3. dependence merging at runtime to reduce the memory overhead of dynamic profiling.

4. dynamic control-flow analysis to record the runtime information of control regions like entry/exit of a function and the number of loop iterations.

DiscoPoP is also able to profile an already parallelized application for dependences [58]. The output of the first phase of DiscoPoP is a CU graph with CUs as vertexes and data/control dependences as edges. It also outputs the control region information including caller/callee and parent/child relationships. The data gathered about control regions during runtime is also reported.

### 1.5.2 Phase 2: Parallelism Discovery and Pattern Identification

Phase 2 of DiscoPoP consist of analyses that discover potential parallelization opportunities in the sequential program using the output of the first phase. DiscoPoP employs different techniques to discover as many parallelization opportunities as possible [59, 57, 60, 61]. These include detection of parallel code regions and parallelizable loops. The identification of patterns for parallel programming is an integral part of this phase and is the scope of this book. A summary of the pattern identification framework is provided in Section 1.6.

### 1.5.3 Phase 3: Ranking

Once the parallelization opportunities have been identified during the second phase of DiscoPoP, these opportunities are ranked according to different metrics [2]. DiscoPoP may suggest more than one parallelization strategies to the programmer. It can be very difficult for the programmer to choose from the available parallelization options. The ranking system of DiscoPoP helps in these situations.

The ranking system exploits different metrics to determine the most promising parallelization opportunity. Instruction coverage calculates the estimated time that will be spent in a code region. It is based on the ratio of instruction count in a region and the total number of instructions in the program. Similarly, local speedup gives an estimated speedup that can be achieved locally if a given code region is parallelized. Lastly, CU imbalance provides the information of load balance between the set of CUs along the critical path of an application. The ranking system computes the global speedup (GS) using the formula:

$$GS = \frac{1}{\sum_i \frac{IC(i)}{LS(i)} + (1 - \sum_i IC(i))}. \tag{1.1}$$

Where $i$ denotes a code region, IC is the instruction coverage, and LS is the local speedup. CU imbalance is used as a tie breaker if the global speedup of two different parallelization opportunities is the same.

## 1.6 Approach Overview and book Contribution

The state of the art discussed in Section 1.4 shows that the automatic parallelization of sequential programs is not an easy problem to solve. This is an active area of research in parallel computing. Jannesari et al. [62] show in a case study of parallelizing real-world applications that the help provided by currently available parallelization tools is not sufficient. These tools need to improve the results produced by them. On the other hand, to ease the burden of parallel programming, software engineers have introduced design patterns for parallel programming. Although these patterns are helpful for programmers, much effort is still needed to find the appropriate choice of patterns to apply them in the software architecture. The complexity of this problem increases manifold if one's job is to parallelize the legacy sequential software. A programmer or software architect needs to have deep knowledge of not only parallel patterns but also the software under study to parallelize it correctly.

Design pattern identification in object-oriented software is an active topic for researchers. Many techniques [63, 64, 65] have been published in this regard. The goal of all these techniques is to identify the underlying design pattern from the UML diagrams of a big software system. It is useful in understanding the design of software in the absence of proper documentation and also helps in further improving the design. Inspired by the same idea, our approach identifies parallel design patterns in a sequential program in order to make the parallelization process easier.

This book extends the parallelism discovery framework DiscoPoP, explained in detail in Section 1.5. The input to our approach is the dependence graph produced by the first phase of DiscoPoP. This book expands the parallelism discovery capabilities of DiscoPoP by identifying multiple parallel design patterns from the algorithm structure of a sequential application. This work is a part of the second phase of DiscoPoP. Our approach automatically infers potential parallel patterns from the input dependence graph and specifies the division of code blocks according to the pattern structure. Based on this information, the programmer can then easily parallelize the code by moving suggested code blocks into appropriate structures of the pattern. This book makes the following contributions:

1. Pipeline and doAll pattern identification using template matching [66]. The template matching technique is widely used in computer graphics. In the case of pipeline identification, we show the mapping of the code sections onto the different stages of the identified pipeline. All the loops are investigated for doAll parallelism and all the loops without inter-iteration dependences are classified as doAll loops. All the cases of inter-iteration dependences are reported. These dependences may be resolved by the programmer to make the loop parallelizable using doAll. We identified pipelines and doAll loops in 12 benchmarks from three different benchmark suites. Our evaluation verified all of these identified opportunities. We achieved an average speedup of $8.4\times$ on a 6 core machine with 12 threads in the best case for a pipeline implementation. We did not find any false positive or false negative cases.

2. Multi-loop pipeline and reduction pattern identification using regression analysis [67] (©2016 IEEE). We analyze the data dependences between two or more loops to identify a multi-loop pipeline. To the best of our knowledge, there is no previous work identifying the multi-loop pipeline pattern. This approach may recommend the fusion of two or more loops if some conditions are met. This analysis also helps to identify reduction when used to detect inter-iteration dependences in a single loop. We implemented the identified multi-loop pipeline pattern in three benchmarks and achieved a speedup of up to $14\times$ on a 16 core machine with 32 threads. Using the same method, we also identify opportunities for geometric decomposition

in a sequential program. The framework identifies a function that has only doAll or reduction candidate loops in it. Such a function is suggested for geometric decomposition. The data passed to such a function can be split into multiple chunks and each chunk can be passed to a separate parallel function call. We identified geometric decomposition opportunities in two different benchmarks. We confirmed our findings by comparing them with the existing parallel implementations provided in the respective benchmark suite.

3. Task identification in a sequential program to exploit master/worker-style task parallelism. We use the classic breadth-first-search to identify different tasks in the dependence graph of a code region as *fork*, *worker*, and *barrier* type tasks. This classification makes the implementation of master/worker pattern easy. We identified the task parallelism pattern in six benchmarks from two different benchmark suites. Our implementations of the parallel versions of these benchmarks achieved a speedup of up to $13.25\times$ on a machine with 16 cores using 32 threads.

The contributions discussed above have been published in peer-reviewed scientific journals or conferences. The next section provides the detailed information about these publications.

## 1.7  Statement of Originality

This book is based on the following peer-reviewd publications, where I have contributed either as the main or as a secondary author:

1. **Using Template Matching to Infer Parallel Design Patterns**, Zia Ul Huda, Ali Jannesari, and Felix Wolf. ACM Transactions on Architecture and Code Optimization, 11(4):64:1–64:21, January 2015. [66]

2. **A profiling tool to identify parallelization opportunities**, Zhen Li, Rohit Atre, Zia Ul-Huda, Ali Jannesari, and Felix Wolf. In Tools for High Performance Computing 2014, pages 37–54. Springer International Publishing, 2015. [9]

3. **Automatic Parallel Pattern Detection in the Algorithm Structure Design Space**, Zia Ul Huda, Rohit Atre, Ali Jannesari, and Felix Wolf. In Proceedings of the 30th IEEE International Parallel & Distributed Processing Symposium (IPDPS), pages 43–52, Chicago, USA, May 2016 [67] (©2016 IEEE).

4. **Unveiling parallelization opportunities in sequential programs**, Zhen Li, Rohit Atre, Zia Ul Huda, Ali Jannesari, and Felix Wolf. In Journal of Systems and Software, 117:282–295, Jul 2016. [60]

5. **Parallelizing audio analysis applications: A case study**, Ali Jannesari, Zia Ul Huda, Rohit Atre, Zhen Li, and Felix Wolf. In Proceedings of the 39th International Conference on Software Engineering: Software Engineering and Education Track, ICSE-SEET '17, pages 57–66, Piscataway, NJ, USA, 2017. IEEE Press. [62]

6. **Dissecting sequential programs for parallelization - an approach based on computational units**, Rohit Atre, Zia Ul Huda, Ali Jannesari, and Felix Wolf. In Proc. of the 10th International Symposium on High-Level Parallel Programming and Applications, Valladolid, Spain, pages 1–18, July 2017. [61]

The work published in all these publications was done under the combined supervision of Prof.

Dr. Felix Wolf (Department of Computer Science, Technical University of Darmstadt) and Prof. Dr. Ali Jannesari (Department of Computer Science, Iowa State University). The contributions of these publications to this book are explained below in detail.

The content of paper 1 in the above list appears mostly verbatim in Chapter 4. I, as the first author, was the main contributor of this publication, with some minor contributions from my co-author, Rohit Atre. I proposed and implemented the idea of using template matching for the identification of parallel patterns. Rohit Atre helped in the evaluation of my implementation.

Similarly, Chapters 5 and 6 are based on paper 3 in the list. Most of the text is taken verbatim from this publication for these chapters. I was the first author and the main contributor of this publication too. In all these two chapters, some additional results have been added that were not part of their respective contributory publications.

Papers 2, 4, and 6 are the ones that deal mainly with DiscoPoP and the detection of computational units in the input program. These papers show how the idea of computational units evolved over time. My contribution to these papers was mainly to the algorithm developed for the detection of computational units. The contents of Section 3.2 are mostly derived from these publications.

Paper 5 in the list provides the case study of different tools and techniques our students used to parallelize different programs during a lab course conducted by our team. I contributed to the supervision of the students and the evaluation of the results together with my fellow co-authors. I reused the results of parallelizing LibVorbis produced in this publication in Section 4.3.1 of this book.

# 2 Design Patterns

It is very rare to encounter an entirely new problem that has never been faced or solved by anyone else when constructing something, such as buildings or software from scratch. Many of the times, the problems faced are either the same or similar to another problem that has been solved earlier. So, reinventing the wheel over and over again is not a good idea as figuring out a good solution for a problem takes time and resources. To overcome this obstacle, the idea of design patterns has been used.

Design patterns provide trusted solutions for recurring problems faced in any field. These solutions are provided by experts in their respective fields and are easily understandable for the non-experts to implement. In this way, design patterns avoid the trouble of going through the whole process of finding appropriate solutions. Design patterns were first introduced for the field of architecture and later adopted by software engineers. In this chapter, we first give a short overview of the usage and advantages of design patterns both in the fields of architecture and software construction. Later, we discuss design patterns for parallel programs in detail. At the end, state-of-the-art techniques to identify design patterns in software are discussed and we summarize the shortfalls of these techniques.

## 2.1 Design Patterns for Architecture

Design patterns provide tested solutions to recurring problems faced in different fields. However, the issue is to record these problems and solutions in such a way that they are understandable to the whole community. To overcome this, Alexander et al. [19] introduced a pattern language. This language bridges the gap between experts providing the solutions and users that want to use these solutions in their projects. The work of Alexander et al. deals with the recurring problems faced in the field of architecture, such as planning and construction of towns, communities, and buildings.

The elements of the pattern language introduced by Alexander et al. are called patterns. Each pattern deals with a recurring architectural problem and provides the basis of its solution so that it can be used whenever the respective problem is faced again. Alexander et al. described each pattern in the following strict format:

1. A figure to visualize the archetypal type of the pattern.

2. An introductory paragraph to explain the pattern context and how is it helpful to fulfill other larger patterns.

3. A headline to give the crux of the problem addressed by the pattern. It is usually only one or two sentences long.

4. A body of the problem to state the empirical background and the different forms of its appearance. The body explains all the details related to the problem.

5. A solution of the stated problem. This solution is stated in a form that tells the user what needs to be done in order to resolve the problem in the stated context.

6. A graphical explanation of the solution, mentioning its different components. This makes the solution easier to understand.

7. A list of other patterns is given in the end which need to be implemented in order to fulfill the needs of the current pattern.

This format provides connections between the patterns in the language. This helps in creating unique combinations of patterns to overcome any problem. Patterns described in the pattern language are in an abstract format and provide the template solution. Each user can modify the solution according to their needs. Alexander et al. also provided comprehensive instructions for the usage of the pattern language [68].

The pattern language for architecture includes a list of 253 patterns. The list starts with the patterns for bigger entities, like towns or cities. After that come patterns for middle-level neighborhoods or buildings. In the end, patterns dealing with construction of smaller units like rooms of a building are provided. This flow helps users to follow through the combination of patterns according to the context of their respective projects and achieve an optimal overall end product.


## 2.2 Design Patterns for Software

The work of Alexander et al. inspired software engineers to create and use software design patterns for their projects. Following the same idea of architectural patterns, software design patterns are introduced to solve recurring design problems faced by software designers. Gamma et al. [3] wrote a book that provide a list of design patterns for object-oriented software systems and their respective description. The boos's authors, namely: Erich Gamma, Richard Helm, Ralph Johnson, and John Vlissides are also know as the Gang of Four (GoF). They provide solutions to the design problems faced in object-oriented programming using classes and objects. A modern-day software engineer is expected to have in-depth knowledge of the design patterns introduced by Gamma et al.

Following the idea of Alexander et al., Gamma et al. also used a pattern language for their design patterns with four main elements:

1. The name of the pattern provides a sense of the problem and its solution, handled by the pattern. A good name of the pattern improves the designing vocabulary and makes communication easier for software engineers.

2. The problem description presents detailed information on the problem and different contexts. This description helps the users to understand better when to use the pattern. Sometimes a list of prerequisites may be given that need to be fulfilled before applying the pattern.

3. The solution contains details of how to solve the problem, in an abstract form. The solution is explained in the form of objects or classes and their responsibilities, relationships, and interactions with each other. The solution gives a template that can be used in different ways to resolve the problem addressed by the pattern according to the different contexts.

4. The consequences describe the effects produced by using the pattern. It includes both benefits and disadvantages of the pattern usage. The consequences help in estimating the cost of using a pattern. They may address the limitations of different programming languages or their impact on the flexibility and portability of the software for using this pattern.

Table 2.1: Elements of a software design pattern. [3]

| Element | Description |
| --- | --- |
| Name and classification | The name of the pattern and class of design issues it addresses. |
| Intent | A short statement describing what the pattern does. |
| Also known as | Any other names of the pattern. |
| Motivation | An outline showing the design problem and how classes and objects solve the problem. |
| Applicability | List of situations where the pattern can be applied. |
| Structure | A graphical representation of the classes in the pattern and the interactions between them. |
| Participants | List of objects and classes in the pattern and their responsibilities. |
| Collaborations | How the participants (see above) collaborate to fulfill respective responsibilities. |
| Consequences | Advantages and disadvantages of using the pattern. |
| Implementation | Any language specific issues or useful hints for implementing the pattern. |
| Sample code | Code segments showing an example implementation of the pattern. |
| Known uses | List of real systems using the pattern. |
| Related patterns | List of other design patterns closely related to the pattern and the differences between them. It also indicates which other patterns should be used together with this pattern. |

Gamma et al. described design patterns as *"descriptions of communicating objects and classes that are customized to solve a general design problem in a particular context"*. They provided a list of 23 classical design patterns. These patterns have been classified according to two different criteria.

The first criterion is the *purpose* of a pattern. The patterns are classified into three categories according to their purpose: *creational, structural,* and *behavioral* patterns. Creational patterns deal with the problem of object construction. Structural patterns provide solutions for the problems relating to the architecture of classes and objects. Behavioral patterns present solutions for interactions between different objects and how they can work together to carry out a task.

The second criterion to classify design patterns is their *scope*. There are only two types of scopes: *classes* and *objects*. All the patterns that deal with the interactions between classes and their subclasses are classified as class patterns. The interaction between the classes is usually inheritance determined at compile time. On the other hand, all the patterns dealing with interactions between objects are classified as object patterns. The objects in a software system are dynamic instances and the relationships between them may change at runtime. Most of the design patterns introduced by

Gamma et al. are classified as object patterns.

Table 2.1 shows a list of all elements of a design pattern and their brief descriptions. These elements provide full detail about the respective pattern and make them understandable for users. Gamma et al. also provided instructions for using design patterns in building software systems. Following their guidance, a user can easily decide on picking an appropriate design pattern for the problem to be addressed. The list of patterns, provided in the related patterns element, helps in determining the group of patterns that can be used to design the whole system. A case study by Gamma et al. provides in-depth details of designing an application using design patterns.

## 2.3 Design Patterns for Parallel Programs

The degree of parallelism available in multi-core processors has increased manifold in the past decade and is continuously on the rise. However, development of software that can exploit the parallel resources available in modern processors is a daunting task. As described in Chapter 1, data and control dependences in a program play a significant role in hampering the parallelization process.

After the huge success of software design patterns in object-oriented programming, software engineers exploited the same concepts for parallel programs [21, 23]. They came up with a comprehensive design pattern language for parallel programs [22]. Keutzer and Mattson [69] have introduced *Our Programming Language* (OPL), that encompasses most of the parallel design patterns. They divide these patterns into five categories, namely: structural, computational, algorithm strategy, implementation, and parallel execution patterns. The work of Mattson et al. [1] has similar impact in the field of parallel programming as the work of Gamma et al. had for object-oriented programs.

Mattson et al. provided a pattern language and a list of design patterns for parallel programming. A programmer having deep knowledge of the problem to be parallelized can follow the pattern language and get either a comprehensive parallel design or working parallelized code. All the design patterns introduced by Mattson et al. are described using the following structure:

1. The **name** of the pattern.

2. A **problem** statement, providing short information of the problem this pattern deals with.

3. The **context** of the problem, stating a detailed background of the problem.

4. A list of **forces**, explaining the main issues that need to be taken care of when designing the solution for the problem.

5. The **solution**, explaining the in detail the usage of this design pattern to deal with the problem. The solution also addresses the issues mentioned in the forces and gives details about how it caters those issues.

6. Some **examples**, showing the user how the design pattern can be applied practically to solve the problem. This makes the design pattern more understandable to users.

7. Some **known uses**, providing a list of real-world programs where this pattern has been used.

Alexander et al. divided their design patterns into categories like patterns for cities, neighborhoods, and rooms. Gamma et al. also divided their design patterns into categories of patterns for classes and objects. Similarly, Mattson et al. classified the patterns they introduced into four design

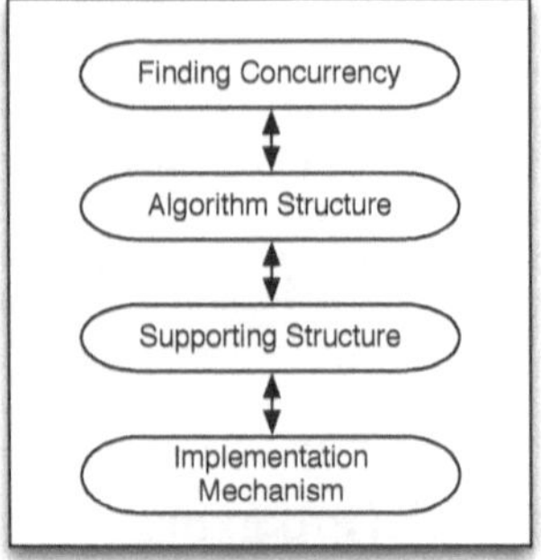



Figure 2.1: Hierarchy of design spaces in pattern language for parallel programs, adopted from Mattson et al. [1].

spaces namely: *finding concurrency, algorithm structure, supporting structure,* and *implementation mechanisms.* Figure 2.1 shows the hierarchy of the four design spaces. The authors also provided a mapping from the patterns of one design space onto the patterns of the next lower space in the hierarchy. This makes the selection of an appropriate design pattern from the next space, based on the pattern from the current one, simple and straightforward. The bidirectional connectors between the design spaces indicate that the user may jump back and forth between design spaces in order to get the optimal set of design patterns.

This book follows the classification provided by Mattson et al. and deals with the design patterns in the algorithm structure as well as supporting structure design spaces. In the following, we summarize all the design spaces and their design patterns in each design space as described by Mattson et al. [1]

### 2.3.1 Finding Concurrency

Designing software always starts with the analysis of the problem to be solved by the software. The aim is to analyze the problem thoroughly and understand all of its aspects as much as possible. Similarly, while developing parallel programs, the problems being solved are needed to be thoroughly analyzed for finding out the best candidate regions to be parallelized. This phase belongs to the finding concurrency design space.

In the finding concurrency design space, the problem is analyzed to discover the work units (tasks) that can execute in parallel to each other. The design patterns in this space help programmers to find such parallelism in the problem space and also guides them on how to arrange the detected tasks to achieve an efficient parallel algorithm. Once, we have the parallelization candidates detected in the problem space, the design patterns in the algorithm structure design space can help in constructing a parallel algorithm. Figure 2.2 shows the design patterns in the finding concurrency design space and their classification.

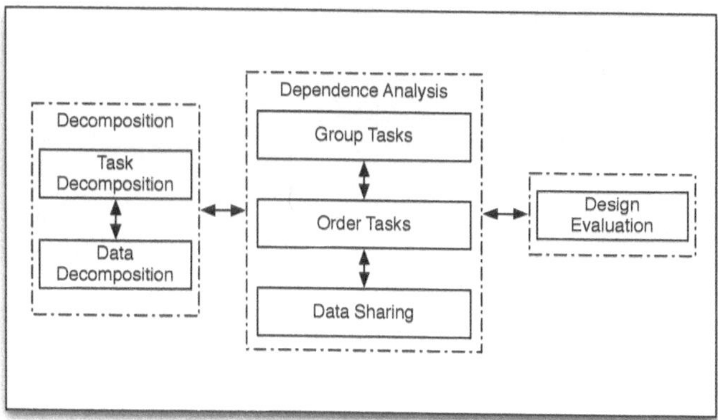

Figure 2.2: Design patterns in the finding concurrency design space, adopted from Mattson et al. [1].

## Decomposition Patterns

After understanding the problem, the first task is to decompose it in such a way that multiple tasks can run in parallel to each other. As discussed in Chapter 1, there are two types of parallelism available in software: functional and data parallelism. The two design patterns in the decomposition subclass help finding these two types of parallelism in the problem.

The task decomposition pattern is applied to the problems where the problem can be divided into more than one independent or loosely dependent tasks consisting of separate instruction streams. The idea is to find tasks with very little dependences among them. This reduces the communication and synchronization overhead of the parallelization. It is recommended to find as many tasks as possible and later merge them to create bigger tasks with reasonable loads. The task decomposition pattern recommends to search for parallel tasks in constructs, such as function calls and the iterations of a loop.

The data decomposition pattern helps in finding the parallelism hidden in the data structures. This implies that the data being used by the program can be divided into independent or semi-independent chunks, which each processing unit can work on, in parallel to each other with little or no synchronization needed. The data decomposition pattern allows programmers to concentrate on parallelism that can be achieved by distributing data and helps in the selection of data structures that can easily support the distribution among parallel tasks.

No matter which type of decomposition pattern is used, the end result is always some parallel tasks either working on their own chunks of data or working in a synchronized manner to address data dependences, for example in a pipeline. The difference between the two patterns is rather artificial, as both present a different view to the same problem. They are explained as separate decomposition patterns for the ease of understanding of the designer. The application of one decomposition pattern implicitly implies the application of the other pattern. A combination of both decompositions can be used to achieve better parallelism in the algorithm of a program.

## Dependence Analysis Patterns

The outcome of the decomposition patterns is a list of tasks that can potentially run in parallel. However, it is not true for all the tasks found in the algorithm. There may be some tasks which are dependent on the output of some other tasks. Hence these dependences are needed to be taken care of. The patterns described in the dependence analysis subclass help achieve this goal.

The group tasks pattern addresses the problem of grouping multiple tasks found in the decomposition phase to make the dependence management simple. This pattern suggests considering two constraints when creating groups of tasks. The first constraint are temporal or data dependences between collections of tasks, meaning that some tasks need data from other tasks. In that case, the dependent tasks need to wait until their input data is available. The other constraint deals with the situation where a set of tasks can be executed in parallel to each other. This usually happens in applications where data decomposition pattern is used. The data is divided into multiple chunks and sometimes computing the boundaries of a chunk needs input from its neighbors. This may cause a deadlock if the neighboring chunk is being computed in parallel. By taking care of these two constraints, the designer can come up with groups of tasks such that the number of dependences are reduced considerably, as they are reported between groups of tasks instead of individual tasks, hence their management becomes simpler.

The solution of the problem of ordering the groups of tasks produced by the group tasks pattern is provided by the order tasks pattern. The goal is to define the execution order of tasks groups in such a way that the semantics of the application being parallelized remains valid. The main constraints for this pattern are the same as that of the group tasks pattern. The main ordering constraint is the existence of a dependence between task groups. A dependent group has to wait for the group producing data for it, hence showing a natural execution order. Some external services of the application may also affect the ordering like an output to a file in a specific order. Some of the task groups may be independent of any constraints and run in parallel in any order. This allows the designers to exploit greater degrees of parallelism in the application.

Once the execution order has been defined for the task groups, the next problem is to handle data sharing between them efficiently. The data sharing pattern addresses this problem. The data being shared between the task groups is divided into three categories. *Read-only* shared data is only read by all the tasks but never updated by any. This type of data sharing can be simply done by just sharing the memory between the tasks. *Effectively-local* data is accessed only by one task and no sharing among other tasks is needed. *Read-write* access to data by more than one parallel tasks is the most challenging type of data sharing. Special care is needed for the order of reads and writes by multiple tasks so that the parallelized application is semantically valid. This can be done using techniques like locks to protect the data being updated concurrently. *Accumulate* and *multiple-read/single-write* are special types of read-write data sharing. In the accumulate type of sharing, the data is used to gather a result like in a reduction. In multiple-read/single-write type of data sharing, the data is usually distributed among the tasks initially, but later on, only a single task reads or updates the data while other tasks just use the initial value. Both these special types of data sharing can be optimized using techniques like reduction and creating a local copy of the data for each task.

**Design Evaluation Pattern**

At the end of the dependence analysis, the software designer has groups of tasks and their order of execution. These groups may communicate with each other using shared data structures. But before continuing into the next design phase, it is always better to evaluate the decisions made during the current design phase and fix any issues found. The design evaluation pattern provides techniques to find problems in the design that has been created so far. If a problem is detected, it can be overcome by going back to the decomposition or dependence analysis steps. This pattern focuses on the three main aspects of the design:

1. The suitability for the target platform is determined. It includes assessment of specific hardware characteristics on which the software will run and how well the design can use the hardware resources available. The issues of number of parallel tasks and efficient data sharing among them are evaluated.

2. The quality of design in terms of simplicity, flexibility, and efficiency is checked. The design should be amended accordingly to create a balanced solution with regards to these three characteristics.

3. The evaluation of the design attributes is done that are important for decisions in the next design space. These attributes include: division of workload among tasks (similar or variable in size), type of communication among tasks (synchronous or asynchronous), and the effective grouping of tasks.

After the careful assessment of all the three issues described above and fixing any bugs or shortcomings in the design, the result is a decomposition of the original problem into subproblems exhibiting concurrency among them. The design patterns in the Finding Concurrency design space work together to create a design that leads to the selection of appropriate patterns in the remaining design spaces. Hence, making the design efficient and bug free in this design space is very important, as a wrong design may lead to wrong solutions in later stages. This would result in the wastage of time and other resources.

## 2.3.2 Algorithm Structure Design Patterns

Once the parallelism opportunities in the problem under study have been identified in the finding concurrency design space, the next logical step is to sketch an algorithm structure that not only exploits the parallelism opportunities found but can be transformed into an efficient parallel program. This is achieved by using the design patterns available in the algorithm structure design space. The input for this space is the design created in the finding-concurrency phase.

To guide through the selection of appropriate patterns from the algorithm structure design space, Mattson et al. provided the decision tree shown in Figure 2.3. This tree is based on the type of decomposition done during the finding-concurrency step and other characteristics determined in the design evaluation pattern.

The first decision is based on the organization of the parallel subproblems done in the previous design space. If the main principal for parallelism is the execution of parallel tasks, then select the *tasks* branch. Similarly, the *data* branch of the tree maybe selected if the decomposition of data is the main source of parallelism. The *flow of data* branch may be selected if the parallelization is

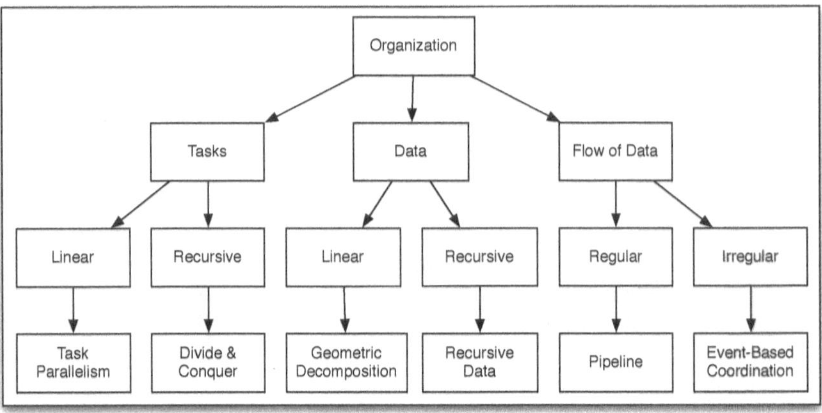

Figure 2.3: A decision tree with all the design patterns in the algorithm structure design space, adopted from Mattson et al. [1].

dependent on the execution order of the tasks.

All the parallel patterns in the leaf level nodes of the decision tree belong to the algorithm structure design space. We briefly describe each of them in the following.

### Task Parallelism

The task parallelism pattern provides the solution for exploiting parallelism if the problem is decomposed into a concurrently running set of tasks. The number of tasks may be known in advance in some cases or they can be generated at runtime. The tasks are usually associated with a program construct like function or loop. Generally, the final result of a program depends on the successful computation of all the tasks in it. But some applications may not need all the tasks to finish before reaching the end goal. A task should be handled and assigned to the processor, satisfying all these constraints. Also, the task scheduler should respect any dependences among tasks, when selecting a task for execution.

The solution provided by the task parallelism pattern is addressed by three main components: the tasks, the dependences among them, and the scheduling of tasks. The tasks defined in the algorithm should ideally outnumber the processors available and the granularity of every task should be big enough to keep the processor busy. The bigger granularity also helps in keeping the synchronization and communication overhead small. The dependences among the tasks should be analyzed and any resolvable dependences (e.g., through transformations or other techniques like reduction) should be fixed. The resolution of dependences among tasks improves the degree of parallelism. Any unresolvable dependences should be addressed properly by applying the Data Sharing pattern.

Lastly, the task scheduler should schedule the execution of parallel tasks efficiently. The main force for efficient scheduling is making the load among the processors as balanced as possible. All the parallel algorithms use one of the two main scheduling techniques: static or dynamic. A static scheduler divides all the tasks into separate fixed-sized chunks and each chunk is assigned to a processor for execution. On the other hand, a dynamic scheduler decides the chunk size at runtime. It is mostly useful if the number of parallel tasks or the number of processors may vary drastically at the runtime.

Often, most of the parallelism found in the programs is in loops. Each iteration of a loop can be considered as a task and multiple tasks can be grouped together into a chunk. These chunks then can be scheduled for execution. If a loop has no inter-iteration dependences, it is also called *embarrassingly parallel* or a *doAll* loop. In some cases, a loop accumulates the results of an operation applied over all the elements of the input data. Such loops can be parallelized by a technique called *reduction*. In a reduction, each task accumulates the results of its chunk of data locally and all these accumulated results are gathered globally in the end.

If the tasks are based on the other constructs, like functions, they can be put into a queue. A scheduler can extract these tasks from the queue and assign them to the processors for execution. The patterns discussed in the Supporting Structure design space provide the appropriate solutions to these problems.

## Divide and Conquer

The dvide & conquer pattern provides the solution for those algorithms that split the problem recursively into subproblems, solve them independently and merge their output in the end to get the complete result. Such algorithms are excellent candidates for parallel execution. With each split of the problem, the number of tasks increases. All these tasks can be executed in parallel and their results can be merged after finishing the execution of all the parallel tasks.

Figure 2.4 illustrates the working of a divide-and-conquer algorithm. With each split of the problem into two subproblems, the number of parallelizable tasks doubles. If the problem can be split into more than two subproblems at each recursive stage, the amount of parallel tasks would increase proportionally to the number of splits. The main constraint of this pattern is to decide when to stop splitting the problem as with each split the amount of work also decreases. This decision can be based on the input data.

The divide & conquer pattern addresses the concerns mostly related to recursive algorithms. Should the newly generated task be directly mapped onto a processor or should it be put into a task queue? This depends heavily on how the work loads are balanced for the generated task. Also, the number of processors and the expected number of tasks play a great role in this decision. The distribution of tasks among processors should also keep a check on communication costs, as the data required by a task has to be moved to the processor where it is being executed. The dependences usually do not cause a problem as normally divide-and-conquer algorithms produce independent tasks. The Data Sharing pattern should be employed in the case of dependences among the tasks.

## Geometric Decomposition

The geometric decomposition pattern focuses on providing the algorithm structure for the problems where data decomposition is the basic driving force for parallelism. The geometric decomposition pattern specifically deals with the problems where chunks of data structures are divisible, like arrays

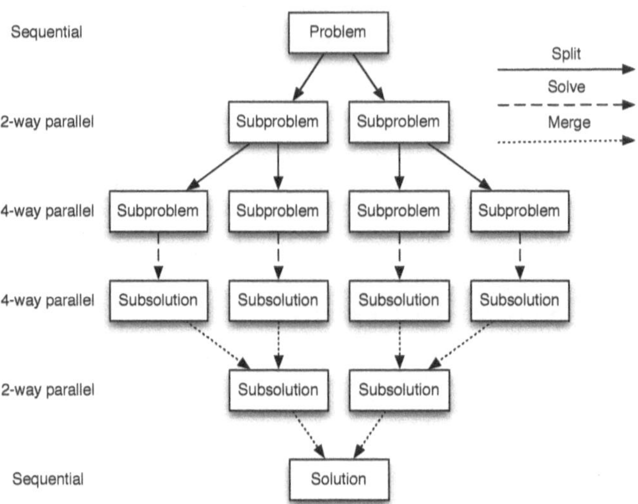



Figure 2.4: Schematic view of a divide-and-conquer algorithm, adopted from Mattson et al. [1].

and are updated concurrently. These data structures are divided into substructures in the same way as a geometric region can be divided into subregions. One task is created for the update of each chunk and all the tasks run in parallel to each other.

In most of the geometrically decomposed problems, updating a single chunk may require the data from its neighboring chunks. The algorithms designed for such a problem should address these data dependences between the neighboring chunks. It implies that the algorithm must ensure that all the data needed by a task to execute properly is already available before its execution. The simplest case is when all the data is available before the start of program. The data can be provided to each task before the start of the parallel execution. But in the other cases, where the data is only available after the partial or complete execution of some tasks, a sophisticated mechanism for communication must be adopted. Mechanisms like locks are very helpful in this regard.

The decomposition of data also needs to take care of the workload for each parallel task. Bad decomposition of data may lead to load imbalance, which in turn can affect performance. Lastly, the data chunks need to be mapped onto processors for actual execution. The same scheduling policies (i.e., static or dynamic) as described earlier, can be used with the same constraints.

**Recursive Data**

Programs working with data structures that are recursive in nature, for example, trees, graphs or lists, require special parallelization algorithms. The recursive data pattern provides solutions for parallelizing such programs. The traversal of a recursive data structure is usually sequential and the parallel algorithm normally requires more computations then the sequential one. Such algorithms are

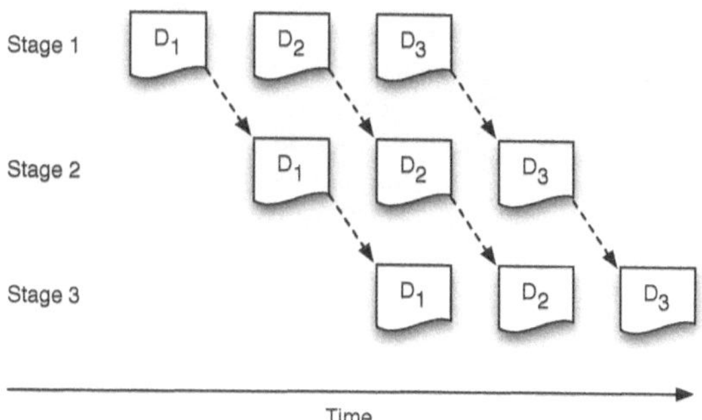



Figure 2.5: Workflow of a pipeline over time. Each stage receives data from the previous stage in the pipeline and performs its computations before passing it to the next stage.

difficult to design as they have to use a different perspective of the problem at hand. Their efficiency depends on the costs of the required extra operations and communications.

The design of an algorithm exposing parallelism based on recursive data structures needs to take care of some key issues. The decomposition of data in these algorithms is always managed at the basic element of the data structure involved, like the node of a tree. Traversing the data structure is usually done sequentially but each element is worked upon individually. The problem arises when there are too many elements, as they can overload the system and result in poor performance. Simultaneous update of the elements also poses a difficulty for the recursive data algorithms. The problem is avoided if the target hardware is an SIMD style machine with lockstep support as it can assign different data to different processors and execute the same instruction on all processors at a time. But the algorithm has to be adjusted if the target hardware does not support this mechanism. The use of separate memory locations for the updated data or synchronization through different mechanisms such as locks become inevitable in these cases.

### Pipeline

The pipeline pattern is useful in cases where the same instruction set is being applied to different chunks of data (like iterating over different elements of an array) and this instruction set can be divided into separate individual groups called stages. To understand the concept easily, Figure 2.5 shows the workflow of a pipeline. Each stage processes the data received from the previous stage in the pipeline and forwards it to the next stage. Hence, each stage is working on a separate data chunk in parallel to other stages of the pipeline.

The pipeline pattern enables parallelization of loops with inter-iteration dependences. It is also called *doAcross* parallelization. Due to these dependence constraints, the stages of a pipeline need to be ordered. The solution proposed in this pattern is simple. It assigns each stage to a separate processor and the data flows between the processors. One stage of a pipeline may have further nested parallelism opportunities that can be exploited separately. Also, sometimes a stage does not depend on the results produced by the same stage in the previous iteration. Such a stage can also be parallelized by concurrently invoking its instances for different iterations. A pipeline with such stages is also known as parallel stage pipeline.

Defining the stages of a pipeline has effects on the performance of the pipeline. The amount of parallelism in a pipeline is limited by its number of stages, but data has to be moved between the stages that creates an overhead. The designer needs to balance the number of stages and the communication overhead. Also the performance is bound by the execution time of the most work-intensive stage of the pipeline. So, a pipeline with load-balanced stages performs the best.

The communication of data among the stages is platform dependent. In message-passing frameworks, this can be done by creating messaging channels between the stages running on different processors. In cases where message passing is not supported, a queue can be used to pass the elements of data between a pair of stages.

**Event-Based Coordination**

The event-based coordination pattern is much like the Pipeline pattern as the driving force is the flow of data through a set of semi-independent tasks. As shown in the decision tree in Figure 2.3, the difference is that in the pipeline pattern, the algorithm structure is linear as well as the communication between the stages is one-directional, regular and loosely synchronous. But the algorithm structure of the event-based coordination pattern is highly irregular. Also, the communication between the tasks is asynchronous and dependent on the occurrence of events as the name of the pattern suggests.

All the tasks in an event-based coordination algorithm have the capability to receive an event, process it and generate new events if needed. Like the pipeline pattern, the events can be passed between the tasks using message passing or maintaining shared queues. The main concern for such an algorithm is to enforce the ordering constraint. For example, a task may not be allowed to process the events in the order they are received. The task may be waiting for a specific event in order to continue its processing. The designed algorithm needs to take care of such restrictions to ensure the correct order of execution. Additionally, it should avoid deadlocks. A deadlock in the program can be avoided by sending required data for an event beforehand. Another technique is the use of special deadlock-detection mechanisms and then resolving the detected deadlocks. Both of these techniques may have extra overhead that can cancel out the parallelization benefits provided by the event-based coordination pattern. Hence, extra caution should be taken to reduce the overhead for handling deadlocks.

### 2.3.3 Supporting Structure Design Patterns

Both of the previous design spaces (i.e., finding concurrency and algorithm structure) deal with the analysis phase of the software design process. Once an algorithm has been designed that exhibits parallelism, its conversion into the parallel code is the next logical step. The design patterns in the supporting structure design space provide solutions for the easy conversion of an algorithm into a program structure. These patterns deal with the details of the parallelism strategies to be used in the program. These strategies can later be transformed into code using implementation mechanisms.

The patterns in the supporting structure design space have been divided into two major subcategories, as shown in Figure 2.6. This division is based on the areas of concern these patterns deal with. The program structure patterns deal with problems of how to exhibit parallelism in the source code. How and when should the parallel tasks be generated and how they should work. On the other

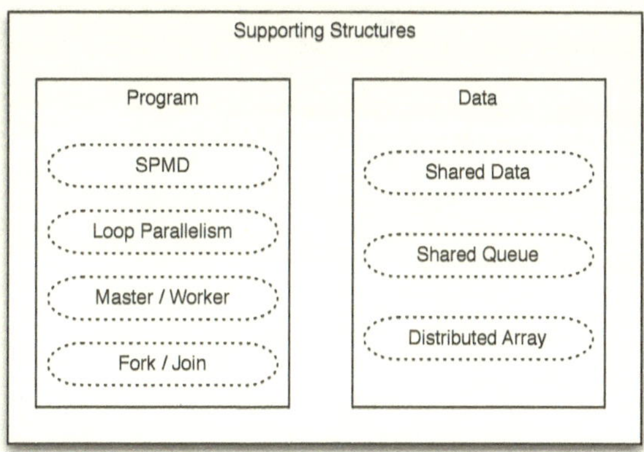



Figure 2.6: Distribution of patterns in the supporting structures design space, adopted from Mattson et al. [1].

hand, the data structure patterns deal with the management of data dependences among the parallel tasks.

All of the patterns in the supporting structure design space help in developing an efficient parallel program. The solutions of these patterns focus on achieving many goals. The resultant parallelism strategy of the parallel program should be easy to understand by just reading the code. It should be scalable and use the available parallel resources efficiently. Easy debugging, verification, and modification of the program is another important aspect. Complex parallel code is hard to maintain and can lead to buggy implementations. It can be avoided if the parallel program's structure is kept simple. Additionally, the choice of programming language and hardware configuration where the program would execute should also be considered when deciding the program structure.

The input of the supporting structure design space is an algorithm designed using patterns from

Table 2.2: Relationship between the patterns of the algorithm structure and the supporting structure design spaces, adopted from Mattson et al. [1].

| Algo / Sup | Task Parallelism | Divide & Conquer | Geometric Decomposition | Recursive Data | Pipeline | Event-Based Coordination |
|---|---|---|---|---|---|---|
| SPMD | ★★★★ | ★★★ | ★★★★ | ★★ | ★★★ | ★★ |
| Loop Parallelism | ★★★★ | ★★ | ★★★ | | | |
| Master/ Worker | ★★★★ | ★★ | ★ | ★ | ★ | ★ |
| Fork/ Join | ★★ | ★★★★ | ★★ | | ★★★★ | ★★★★ |

the Algorithm Structure design space. A major problem for a designer is to choose an appropriate pattern from the Supporting Structure design space for this algorithm. Mattson et al. provided a mapping of the patterns from the two design spaces, as shown in Table 2.2. This mapping implies that a pattern of the Supporting Structure design space maybe helpful in implementing more than one patterns from the Algorithm Structure design space. The selection of the pattern is up to the designer based on the program needs and facilities provided by the pattern. The parallelism suggestions by our parallel pattern identification approach closely follow this mapping. We summarize features of the supporting structure patterns in the following.

**SPMD**

A dominating number of parallel programs exhibit single program multiple data (SPMD) type of parallelism. The same program constructs (e.g., loops and functions) are executed on different units of data in parallel in such programs. The control flow in these constructs may vary due to some specific conditions based on the set of data received as input and the processor id of the executing processor. So, the source code for most of the parallelized regions is almost always the same. This makes the source code abstract and easily manageable. Due to these benefits, the SPMD pattern is the most used pattern to structure parallel programs.

The parallel code structure developed using the SPMD patternshould attend to some of the problems that may arise. The SPMD pattern uses similar code for all parallel tasks but sometimes applications may need to run a totally different set of instructions on different types of data, hence adding complications. The program created should be portable and use the most commonly available hardware features.

The template solution provided in this pattern has been divided into five stages. The *initialization* phase consists of loading the parallel program onto a processor and setting up all the communication channels to be used by the parallel subprocesses or threads later on. The next phase is to *assign a unique identifier* to each thread. This identifier may be used later to determine the behavior and workload of the thread. The third phase of *running the same program on each thread* uses the unique thread identifiers to decide different execution paths. Usually, branching instructions or the division of the loop iteration space based on this unique identifier achieve this goal. The fourth stage is to *distribute the data* according to the thread identifiers. This is done by either moving the relevant data of each thread to its own local storage space or sharing the whole data structure, such that each thread uses a subset of data based on its identifier. Lastly, the *finalization* stage closes the parallel code section and terminates all threads. Data may be gathered before termination if needed and merged to get the complete result.

The flexible programming structure of the SPMD pattern is compatible with all six patterns of the Algorithm Structure design space, as can be seen in Table 2.2. The function of data distribution based on unique identifiers should be simple enough to make the structure more understandable. Copying whole data for each subprocess can create big overhead and should be avoided. In the SPMD pattern, the distribution of data and the set of parallel instructions is determined statically, hence avoiding any scheduling overhead.

**Master / Worker**

As described earlier, the SPMD pattern distributes tasks and data statically. A problem arises if the tasks have variable load. This load imbalance can lead to a very inefficient parallel execution. The master / worker pattern provides a solution for this problem. The basics of this pattern are similar to that of the SPMD pattern, except the work distribution is done dynamically, instead of statically based on unique identifiers.

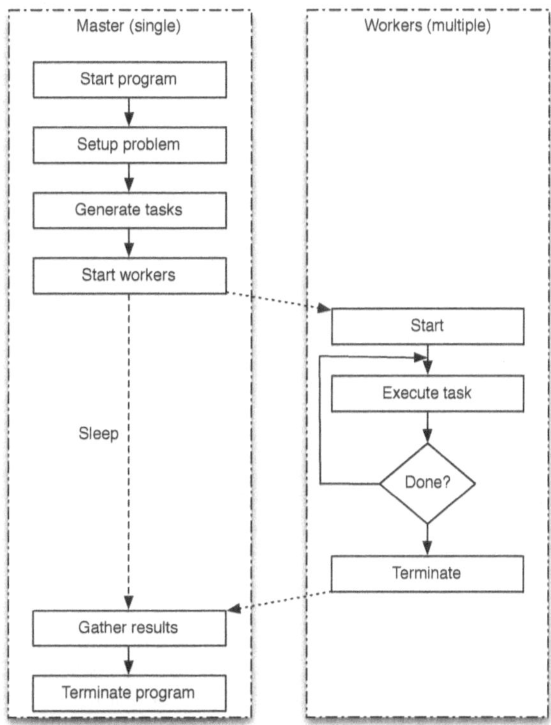


Figure 2.7: The usual workflow of the master / worker pattern, adopted from Mattson et al. [1].

The master / worker pattern is suitable for loosely coupled parallel tasks with little to no communication between them. This pattern handles the load imbalance among the tasks automatically. A programmer may try to divide the larger tasks into smaller ones to balance their loads, but this may lead to bigger communication overhead and make the program error prone. Therefore, the pattern suggests to schedule the smaller tasks around the bigger ones.

A typical workflow of the template solution proposed by the master / worker pattern is shown in Figure 2.7. A single master thread starts the parallel program and initializes all the required resources. Before the start of the parallel region, the master creates a set of tasks to be executed in parallel and launches multiple worker threads to execute these tasks. Each worker thread gets a task out of the task set and executes it. At the end of task execution, the worker checks the task set for a new task and executes it if available. All workers terminate once the task set is empty. The master gathers all the results in the end and terminates the program. This strategy automatically handles load imbalance as each worker may receive a new task as soon as it finishes the assigned task.

The solution proposed by the master / worker pattern scales well if the number of tasks exceeds the number of workers and the workload is much greater than the communication and synchronization overhead imposed by the solution. The tasks set requires a global communication mechanism among the workers and access to it should be optimized. The Master / Worker pattern is the best candidate for implementing the Task Parallelism pattern from the Algorithm Structure design space. Implementing other patterns from the Algorithm Structure design space with the master / worker pattern needs more modifications to the typical solution and may make the source code very complex.

**Loop Parallelism**

Most of the scientific applications have computation-intensive loops. The parallelization of legacy code written for such applications requires these loops to be parallelized. The loop parallelism pattern provides a solution to this type of parallelization process.

This pattern is applied in cases where a sequential code already exists and implementing the parallel version from scratch is not a viable option. Such applications are usually very complex and algorithms are not clearly understandable by just reading the code. As most of the scientific applications require precision in results, the parallelized code developed after applying the Loop Parallelism pattern should have sequential equivalence. To achieve this goal, the code is usually parallelized incrementally (one loop at a time) and the application is tested for the sequential equivalence after each step. The programmer may need to restructure the loop code for the efficient use of the memory hierarchy to achieve better performance.

The solution listed in this pattern has been divided into four steps. Initially, loops with most of the workload of the application need to be identified. Such loops are also called hotspot loops. They can be identified by looking at the code or using application profiling tools. The next step is to resolve any loop-carried dependences, such as a RAW dependence between two difference iterations of the loop. This can be achieved by using different transformations or using different synchronization mechanisms. The third step of the process is to parallelize the loop by splitting up iterations among different parallel tasks. This may also require dividing data among different tasks. The fourth and last step is to tweak the parallelization schedule to get optimal performance from the parallel code. Both static or dynamic schedulers may be tested with different chunk sizes to achieve this goal.

The pattern suggests two optimizations that can be used to further improve the parallel version. *Loop Merging* can be used if many loops iterate over the same range. Such loops may be merged into a single larger loop. This is also known as *Loop Fusion*. The second technique is called *Coalescing* of nested loops. This technique combines nested loops into a single larger loop. Both of these techniques avoid parallelism overhead by reducing the number of loops to be parallelized and making the parallel tasks more coarse-grained. The most appropriate API for parallelizing applications that fall into the category of the Loop Parallelism pattern is OpenMP. It supports incremental parallelization and the changes required to parallelize the sequential code are minimal.

## Fork / Join

Some applications have parallel tasks generated at different stages of their execution and the tasks generated at a single stage have a strict relationship with regards to each other. Parallelization of such applications cannot be handled by simple task or loop constructs. The fork / join pattern presents solution for this type of problem. The parallel tasks are forked at runtime and later merged back into the forking task.

The algorithms using the Divide & Conquer pattern from the Algorithm Structure space are the best candidates for the fork / join pattern. When a task requires subtasks, it forks new parallel tasks and idles itself. Once all of the subtasks are executed, they report their results to the parent parent task and finish their execution. The parent task resumes its execution after all the forked tasks have finished. The pattern works best for the algorithms where tasks are bound by a complex relation or are recursive in nature, for example merge sort. Special care needs to be taken in regards to making the solution load balanced.

The solution listed in the fork / join pattern mainly focuses on two different techniques. The first approach is the simple one. It creates a new processing thread for the execution of each new task being created. Most of the parallel programming frameworks support this type of parallelization. At the end of the execution of the task, the thread also terminates. However, this approach may not be efficient if the creation and destruction of tasks is too frequent, as it increases the overhead for thread creation and destruction, which are very expensive operations. To overcome this disadvantage, the pattern suggests the other approach where a team of threads is created only once and new tasks are stored into a work queue. Any thread that finishes its task may get a new task from the task queue. This approach is more like that of the master / worker pattern with the difference that in this solution, any worker thread may also produce new tasks to be executed.

## Data Structure Patterns

All the above-mentioned patterns are specific for program structures. Most of the patterns in the algorithm structure space implicitly resolve data dependences using techniques like reduction in task parallelism or synchronization among the stages of a pipeline. But in some applications, these solutions do not work out of the box and data sharing is needed among the parallel tasks. Hence, supporting structures are needed to take care of the shared data. The shared data pattern from the data branch of the supporting structures design space deals with these problems.

The major concern of the shared data pattern is to produce a solution that ensures correct results of the computations for any execution order of the parallel tasks. The data-management overhead should be kept small and the structure should be able to scale with the availability of more parallelism. The solution of the Shared Data pattern needs to be easily understandable. A complex data structure may lead to concurrency bugs and is difficult to maintain over a longer period of time.

Implementing a separate solution to handle shared data adds to the complexity of the parallelized program. The pattern recommends to avoid the shared data structures if possible and rethink the software design. If the shared data structure is unavoidable, then it should be designed on an abstract level. A definite set of operations should be defined with well explained details of their functionalities, such as the push and pop operations of a queue. The next step is to identify the set of operations that do not interfere with each other if executed in parallel, for example, updating of different data objects. The execution of the rest of the operations should be controlled using either locking or lock-free mechanisms so that they do not interfere with each other.

The shared queue and distributed arrays patterns are specialized versions of the shared data pattern. A shared queue works mostly like a normal queue and provides the standard operations like push, pop, and isEmpty etc. The internal mechanism of the shared queue ensures that no two pop functions return the same data object and also two concurrent push functions result in the correct insertion of data. Similarly, the distributed arrays pattern provides structures to share data in consecutive memory locations. The structure divides the data units into chunks and provides concurrent access to each chunk. However, some control mechanisms may be implemented for accessing neighboring chunks or data elements at the boundaries of the chunks.

### 2.3.4 Implementation Mechanisms

After the application of appropriate patterns from the three design spaces discussed so far, the software design achieved consists mostly of the high-level constructs. This software design needs to be translated into the source code using a programming language. Almost all of the programming languages have one or more parallel programming interfaces. Some of the basic low-level features are similar in all of these parallelism frameworks. The *Implementation Mechanisms* design space deals with all such features. These features have been divided into three main categories: thread / process management, synchronization, and communication.

Thread or process management includes creation, destruction, and usage of threads. Each programming framework has its own mechanism of creating and destroying threads. Some support explicit creation and destruction by calling specific functions during the execution. Others do not provide such explicit access to threads and hide these operations from the programmer. The programmer just needs to provide information about the parallel regions in the code and the rest is handled by the framework internally. Any of these two approaches can be used by the programmer based on the requirements of the program as well as the capabilities of the framework being used.

Synchronization of parallel tasks is an important aspect of parallel programming. The synchronization among the tasks is usually required to keep the order of operations on some shared resources. All the frameworks provide some synchronization mechanisms. A thread may be waiting for a specific value of a variable to start its execution. Techniques like memory fences make sure that the data is consistent across all the threads. This is achieved by using special directives made available by frameworks like `flush` in OpenMP. Barriers are used to synchronize all the threads active in a program. No thread is allowed to continue further execution unless all of the threads in a team reach the barrier point. This mechanism helps achieve a consistent view of all the resources for all the threads. Another technique called mutual exclusion ensures that a part of the code is executable only by a single thread at a time. Such a piece of code is also known as critical section. This mechanism helps overcome data races in parallel programs.

The third category of the Implementation Mechanisms design space is communication among parallel tasks. Different frameworks achieve this by using different techniques. Some pass messages among parallel tasks. These messages may be directed to a specific task or may be sent to the whole set of tasks. Each task receives the message sent and uses the data enclosed accordingly. On the other hand, some frameworks that are specific for multi-core systems take benefit of the common memory available to all the processing cores and copy the data in a shared location. This data can be accessed by the tasks that require it. Of course, some kind of notification system is required in this case to alert the receiving tasks about the availability of data.

We have given an overview of the four design spaces available in the parallel pattern language introduced by Mattson et al. Use of this pattern language requires in-depth knowledge of the problem under analysis and the patterns available in all the four design spaces. On the other hand, the process can be eased if the sequential program is present and locations can be identified in it where any of these parallel patterns can be applied.

## 2.4 Identification of Sequential Design Patterns

Object-oriented software design patterns make a software system easy to understand and manage. If the design is correctly documented during all of the software engineering phases, the process of maintaining and changing the software becomes simplified. A software engineer can bring up the required changes by keeping in mind the software patterns already in use, so that the design based on them is not broken. This also prevents the introduction of new bugs and makes the testing process easier. However, if the software system is complex and there is no up-to-date documentation available, making even a small change in the design can become a challenging task.

Reverse software engineering is the branch of software engineering to deal with the problems where a complex software system exists but in-depth knowledge of its design is lacking. In the reverse engineering process, the design of the current system is analyzed to identify any occurrences of software design patterns. As the software system under examination may be complex in nature, the pattern identification process needs to be automatic. A lot of research has been done in this regard. In the following, we describe some of the state-of-the-art approaches for the automatic identification of object-oriented software patterns in existing applications.

The *Unified Modeling Language* (UML) [70] is the standard language to design software. All of the software design patterns come with their respective standard UML diagrams. These diagrams are the base for all of the pattern identification approaches. Gupta et al. [71] convert the the UML diagrams of each pattern introduced by Gamma et al. [3] into separate directed graphs. The vertices of these graphs are the classes used in the pattern and the edges are the relationships among the classes in the pattern like generalization, direct association, etc. Similarly, the UML diagram of the application under study is also converted into a directed graph. All these directed graphs are then translated into boolean functions in the *sum of product* (SOP) form. Once the SOP form is established, Gupta et al. use a substring search algorithm to identify patterns in the SOP of the application. The result is the complete or partial identification of a pattern in the application. The drawback of this approach is that it cannot be used for identifying the patterns containing classes with self-associations.

Tsantalis et al. [65] propose using a similarity scoring algorithm designed for detecting similarity among matrices. For this purpose, they convert UML diagrams of both the patterns and application into directed graphs similar to Gupta et al.'s approach. The adjacency matrix representations of these graphs are used as inputs to the similarity scoring algorithm. The output of the algorithm is in the range $[0 - 1]$. A 1 indicates matrices are similar, hence the pattern exists. A 0 indicates no similarity between the matrices and the pattern does not exist. A value in between 0 and 1 indicates the partial existence of the pattern. This approach shows the successful identification of patterns in the application, however the algorithm can be very time consuming if the application graph is too big. There are many false negatives reported due to slight change in the implementation of a pattern in the source code of the application. Also, this approach fails to correctly identify a pattern if there are multiple instances of it in the application.

A template-matching based pattern identification technique is introduced by Dong et al. [72]. Template matching is a well known technique used in the computer vision field. It uses cross correlation to detect similarity among two different vectors. Dong et al. improved the technique of Tsantalis et al. by converting the adjacency matrices of the directed graphs into vectors and computing the cross-correlation between them. The results of cross-correlation are interpreted in the same way as that of similarity scoring. A result of 1 means similar vectors, meaning the pattern exists fully, while a result of 0 means the absence of the pattern, as the vectors do not match at all. Similarly, values in between zero and one mean partial existence. Dong et al.'s method overcomes the drawbacks of Tsantalis et al.'s approach for the identification of multiple instances of a pattern in an application.

Qiu et al. [73] also use the UML diagrams of both the application and the patterns. They convert these diagrams into abstract syntax graphs in which the vertices are abstract classes, classes, or methods and edges are associations among the vertices. Qiu et al. discover subgraphs via graph isomorphism. They developed finite space machines called *state space graphs* (SSGs). The end states of these SSGs always match the model of a specific design pattern. Qiu et al. pick vertices and edges from the application's diagram that match the initial state of an SSG. Further edges and vertices are added according to the next steps of the SSG if possible. The process continues until the final stage of the SSG is reached, meaning successful identification of the pattern, or until a threshold is reached resulting into an unsuccessful attempt. The evaluation by Qiu et al. shows that they successfully identified eleven different patterns from three complex Java-based applications. They claim to obtain better results in comparison to Tsantalis et al. However, their approach still reports false positives.

Hayashi et al. [64] identify meta-patterns [74] in the underlying applications. Meta-patterns deals with the meta-information of a program. Hayashi et al. use both static and dynamic analysis to extract information from the application. All this information is converted into Prolog facts. Also the specifics of meta-patterns are converted into Prolog rules. The Prolog interpreter uses these rules to identify meta-patterns from the application. Maggioni and Arcelli [75] use a metrics-based identification process for micro-patterns [76]. Micro-patterns are more related to the lower level implementation details. Maggioni and Arcelli use the number of methods and the number of attributes in a class as metrics to identify the patterns in the application's source code.

## 2.5 Identification of Parallel Design Patterns

The identification of design pattern in object-oriented sequential programs provides many advantages, as discussed in the previous section. Similarly, the identification of parallel patterns can give insights to the programmers to improve their parallelization. If it is done for already parallelized applications, it can help users to take informed decisions about the application's runtime environment and get optimal performance. Also, different bottlenecks may be revealed and mitigated accordingly. On the other hand, parallel pattern identification in sequential applications reveals different parallelization opportunities. This can help a programmer tremendously in the parallelization process. Below, we discuss some state-of-the-art techniques and tools introduced in the literature for parallel-pattern identification.

### 2.5.1 State of the Art

The identification of parallel design patterns in already parallelized applications is an active field of research. Poovey et al. [77] identify parallel patterns in a parallelized application using a mix of different metrics. They have categorized data sharing patterns among the threads into four categories: read-only, migratory, producer / consumer, and private. All of these categories imply the presence of an algorithm structure space pattern to some extent. Poovey et al. also monitor different events in the application like the creation and destruction of threads, thread imbalance determined by the number of instructions per thread, and the uniqueness of the program counter of each thread. They use an online detection mechanism for pattern identification. The information extracted by this approach can be used to enhance the performance of the parallel application by improving thread scheduling, load balancing, and mitigate hardware-specific problems.

MINIME [78] is a tool that leverages the same technique as of Poovey et al. to search for parallel patterns in parallel applications. However, its ultimate objective is to produce a synthetic benchmark with the same parallel patterns and parallelization behavior. These benchmarks can then be used for testing a hardware architecture under development. Mazaheri et al. [79] analyze communication among the threads of a parallel program to identify different communication patterns. They use DiscoPoP [2, 9, 10] to instrument the parallel application and count the instances where a thread reads the data written by another thread. Their results can be used for improving performance by optimizing thread mapping for each parallel region.

Identification of parallel patterns in sequential applications helps in the parallelization process. Jahr et al. [79] provide a manual approach to identify parallel patterns in sequential embedded-system applications. They use the UML activity diagram of a sequential application to find appropriate locations in it where parallel patterns can be applied to achieve parallelism. However, their approach needs comprehensive knowledge of parallel constructs and patterns. To overcome this problem, many tools have been developed to automatically identify parallel patterns. We discuss some of them in the following.

#### Decoupled Software Pipelining

*Decoupled Software Pipelining* (DSWP) was introduced by Rangan et al. [80]. This technique parallelizes loops with recursive data structures, like linked lists, by converting them into pipelines. DSWP splits the code into two stages of a pipeline. The first stage is called a producer stage. It iterates over the recursive data structure to extract pointers of the data elements. These pointers are then passed over to the second stage called the consumer stage. This stage retrieves the data pointed by the received pointer and performs computations. DSWP does not perform well when executed on commonly available processors. This is due to the lack of an effective data flow mechanisms among the cores of these processors. To overcome this problem, Rangan et al. propose a special processor with synchronized arrays implemented in hardware. These arrays facilitate efficient communication between the cores executing producer and consumer stages. The evaluation of DSWP shows an improvement of $11\% - 76\%$ in runtime over the sequential versions of programs.

Approach by Rangan et al. works only for loops with recursive data structures. To overcome this limitation, Ottoni et al. [81] extended DSWP to identify pipeline patterns in general loops with inter-iteration dependences. Their approach identifies the Pipeline pattern in loops and transforms them automatically to implement the identified pipelines. The DSWP compiler statically builds a

*Program Dependence Graph* (PDG). This graph contains both data and control dependences and uses basic blocks as its vertices. They apply the *strongly connected components* (SCCs) algorithm from graph theory to the PDG of a hotspot loop to reduce it to a directed acyclic graph. This graph forms the basis of a pipeline. The extended DSWP is developed only for dual core processors with a special hardware-based message passing mechanism available between the two cores. So the computed directed acyclic graph is split into two subgraphs with each becoming a stage of the pipeline.

The evaluation reported by Ottoni et al. presents promising results for different general purpose sequential applications. The best speedup they achieved is 145%. Raman et al. [82] improved the DSWP algorithm by splitting the directed acyclic graph generated by the SCC algorithm into multiple stages, compared to only two stages in the original DSWP. This improvement resulted in the identification of pipelines with parallelizable stages. All the stages that do not have any inter-iteration dependences can be run in parallel to each other, hence improving the performance of the implemented pipeline. The parallel-stage DSWP showed better scalability than the original DSWP with a speedup of up to 155%.

Vachharajani et al. [83] further extended DSWP to include speculative parallelism. They speculatively executed the stages of a pipeline and rolled back the progress in the case of a wrong speculation. Huang et al. [84] proved that the stages of the pipeline identified by Raman et al. can have more parallelism after analyzing the dependences of each stage separately. DOMORE [85] uses DSWP to generate code for one master and multiple worker threads. It also detects dependences between iterations at runtime and synchronizes the iterations if required.

DSWP provides the basis for many parallel pattern identification approaches. However, there are some drawbacks of DSWP. It does not result in any significant speedup when used on commonly available processors. It requires specialized communication arrays or message-passing channels among the producer and consumer processing cores. Hence, the evaluation is done on a simulator as the required hardware is not available. Additionally, all the limitations of static analysis, like branch explosion and pointer aliasing etc., may affect DSWP, too.

### Sambamba

Sambamba [86] is an automatic parallelization framework. It identifies multiple parallelization schemes for each function of the application. At runtime the best combination of these schemes are selected according to the execution environment. These schemes can also be changed automatically if the execution environment changes. The compiler of Sambamba analyzes the input program and transforms it accordingly. It also performs scheduling. The runtime component of Sambamba optimizes the application during execution using static information as well as dynamically profiled data.

The Sambamba compiler constructs a PDG of the application under analysis. The vertices of this PDG are basic blocks, while the edges are the data and control dependences computed statically. It also uses hints provided by the programmer for dependence analysis. Sambamba uses already available techniques for parallelism discovery. It covers a wide range of parallelism techniques, including: doAll, doAcross, reduction, task parallelism, privatization, and speculation. The Sambamba scheduler uses an approach based on integer linear programming (ILP) to identify different parallelization opportunities. An execution cost model is computed based on the information statically gathered from the application. At runtime, the statically identified parallelization candidates are evaluated for the available runtime environment. A dynamic dispatch system decides at runtime whether the application should continue using the selected parallel scheme or whether it should be changed to another scheme with the help of a just-in-time compiler.

The experimental evaluation shows that Sambamba outperformed automatic parallelization performed by Polly [4]. This is because Polly misses some parallelization opportunities that are identified by Sambamba. The use of static analysis for parallelism discovery restricts Sambamba to a single function in the code. This limitation can lead to some missed parallelization opportunities. Also the data dependence analysis technique used by Sambamba cannot cover data structures such as arrays or the effects of recursive functions.

## HELIX

HELIX [87] is a fully automatic parallelization technique for embedded system applications. It is used to parallelize loops with inter-iteration dependences using wait mechanisms among threads. The parallelized version of the sequential input application is a special type of pipeline called homogeneous pipeline [88]. The idea is to run one iteration of a loop per core. The inter-iteration dependences are handled by employing *wait* and *signal*. Each consumer iteration waits for the data to be available and each producer iteration signals the consumer once it generates the required data.

HELIX exploits many sophisticated techniques to reduce inter-processor communication caused by inter-iteration dependences. It only parallelizes the loops selected through a complex cost model and tries to minimize the segments of an iteration causing inter-iteration dependences. The iterations are assigned to processors in a round robin fashion and each processor can only read a signal from its own memory space and write a signal in the memory space of the next processor. This simplifies the signaling logic and reduces synchronization overhead. The maximum reported speedup of HELIX is $4.12\times$ for six processors.

HELIX-RC [89] suggests an improvement of HELIX and proposes a new hardware component called ring cache. A ring cache exists between adjacent processor cores and data flow is unidirectional. As HELIX assigns the iterations of a loop to each core in round robin fashion, the ring cache can achieve maximum parallel efficiency by reducing communication latency. HELIX-RC also suggests decoupling the signaling mechanism from the actual computations by adding special instructions to the instruction set architecture. This further reduces the synchronization overhead of HELIX.

HELIX-RC achieves a maximum speedup of $12\times$ on a 16 core machine. HELIX and HELIX-RC are designed for embedded system applications and they parallelize loops, which are thought to be difficult to parallelize, successfully. However, they are limited to irregular types of loops and the signaling can become a bottleneck if there are too many inter-iteration dependences. Also, the communication overhead can increase many-fold if the amount of data to be transfered among the iterations gets too big.

## PARAMAT

Keßler developed an automatic parallelization tool named PARAMAT (PARallelize Automatically by pattern MATching) [90]. PARAMAT is specifically designed for scientific applications such as simulations. Keßler investigated a bulk of scientific applications and came up with around 150 patterns for these applications. These patterns deal with nested loops. PARAMAT's basic pattern library contains all these patterns. These patterns are classified into a hierarchy of five different levels. Each level denotes the depth of a loop nest, indicating the depth level at which that pattern may exist. This classification also results into a natural hierarchy of the patterns.

The pattern recognition algorithm of PARAMAT uses the abstract syntax tree computed from the source code. It starts recognizing patterns for each leaf node of the tree and continues in upward direction. For each node in the tree, the pattern is recognized according to the patterns already matched for its children. The hierarchy of patterns accelerates the search. Hence, pattern recognition can also be seen as a path finding problem. For this reason, the pattern recognition algorithm of PARAMAT has very efficient runtime.

Once a pattern has been identified, the code is transformed using the templates available in the PARAMAT library for that pattern to parallelize the code. These templates provide abstract parallelized solutions for their respected pattern. They may include a very basic parallelization scheme or highly optimized hardware-specific solutions. The use of a library for both patterns and their templates make PARAMAT easy to extend. The pattern matching algorithm and transformation mechanism of PARAMART is very efficient and robust, but it only supports the parallelization of scientific applications and is not applicable to general purpose applications such as text processing or spreadsheets.

### Profile-Based Pipeline Identification

A framework to identify pipeline parallelism in general purpose sequential applications is introduced by Rul et al. [88]. Boundaries of all regions like functions, loops, and basic blocks from the input program are recorded statically. The program is instrumented for data and control dependence profiling. This framework works more like our approach. The main difference is that the profiler developed by Rul et al. records data dependences for the whole data object instead of a single element of the object. This reduces the profiling overhead but may lose some important information.

The dependences are recorded by executing the profiled program and later analyzed for pipeline identification. The post-analysis builds a data dependence graph with code regions and data objects as its nodes. The edges of this data dependence graph are the data dependences between the regions and the data objects of the program. Loops with inter-iteration dependences and function calls in them are selected for possible pipeline identification. The subgraph of each function node in the graph of a selected loop is clustered into a single node and the algorithm tries to find a data-flow path from the first node until the last node of the graph. The successful detection of a data-flow path results in pipeline identification.

The framework automatically transforms sequential into parallel code according to the identified parallelization opportunity. Rul et al. reported the successful identification of pipeline and doAll parallel patterns. They tested the parallelization framework on different benchmark suites and reported a 40% speedup.

### Others

Numerous other researchers have worked on parallelism identification. Astorga et al. [91] identify pipelines in the abstract syntax tree of a sequential application. They use static analysis for this purpose. Molitorisz [92] proposes a tool, which identifies parallel design patterns and applies automatic transformations to create a parallel version of the serial input application. He is able to identify master/worker and pipeline patterns. He tracks control dependences to identify these patterns, mainly focusing on dependences between function calls. Cordes et al. [93] identify task parallelism in a hierarchical task graph for heterogeneous architectures. The tasks identified by their approach can be at the instruction level, the loop level, or the function level, depending on the available architecture.

Lee et al. [94] extend the Cilk programming interface to add annotations that support on-the-fly pipeline parallelism. They added a new set of keywords to Cilk. A programmer annotates the program with these new Cilk annotations after which it is compiled into a parallel application. The PIPER scheduler developed by Lee et al. uses work stealing to orchestrate the identified pipelines. Unnikrishnan et al. [95] developed a technique to efficiently implement doAcross parallelization. They fold all of the inter-iteration dependences of a loop into a single, conservative dependence. This reduces the communication overhead when iterations of a loop are run on different threads. Their approach uses wait and signal to synchronize threads. They try to move all the instructions generating data for the inter-iteration dependences to the earliest possible position in the execution order and all the instructions consuming this data late in the execution order to further improve the efficiency of the parallel application. Zhang et al. [96] discuss a technique to further enhance the performance of the pipeline pattern by executing multiple iterations as a single chunk per stage instead of one iteration at a time. Their evaluation shows significant performance gains.

### 2.5.2 Comparison with our Approach

In the previous section, many state-of-the-art approaches that automatically identify parallel design patterns have been discussed. The following points differentiate our work from them.

- Many of these approaches use only static dependence analysis. As discussed in Chapter 1, static analysis has the advantage of being fast and covering the whole code base. But analyzing dependence statically has many disadvantages, too. The use of pointers and complex or dynamic array indexing hinders data-dependence detection. Hence, the applications using these techniques do not gain much from such methods. We use a hybrid approach that combines the advantages of both static and dynamic analysis to overcome these hurdles.

- Most of the methods identify at most one or two different types of parallel patterns. All such approaches may overlook patterns they do not cover. On the contrary, our framework identifies almost the full spectrum of parallel patterns available in the algorithm structure design space and provides an appropriate mapping according to the solutions listed in the supporting structure design space. Hence, our approach has high a probability of finding the appropriate parallelization pattern.

- Some of the techniques like DSWP and HELIX need special hardware components for efficient parallelization. Building a new type of hardware is a very expensive task. It is not economically viable unless the applications being addressed are frequent and commonly used. The parallelization opportunities identifies by our framework can be implemented on any hardware.

- Many approaches are developed for specific types of applications, such as PARAMAT for scientific applications or HELIX for embedded systems. Our pattern-identification framework is not limited to any specific type. We focus on general applications and our solution can be used for all types of applications.

The automatic transformation of sequential code into equivalent parallel code is an important issue in parallelization. Many state-of-the-art tools offer this capability with limitations. The ultimate goal of our framework is to also achieve automatic transformation by overcoming most of these limitations. Li et al. [97] attempted to accomplish this task but did not achieve best results. On the other hand, Norouzi et al. [98] used our pattern identification framework to automatically map the identified patterns to the constructs of OpenMP. They showed very good results compared to the other state-of-the-art automaic parallelization tools. Still, we need to work out a more effective solution for automatic parallelization in the near future. Therefore, it is beyond the scope of this book.

## 2.6 Summary

A design pattern provides tested solutions to a specific problem encountered by designers and engineers. Design patterns were first introduced for the field of architecture in the 1970s. Christopher Alexander provided a catalog of design patterns for building cities, communities, and buildings. Software engineers followed the idea of design patterns and introduced software design patterns. Especially, the patterns by Gamma et al. became well known in the field of software engineering. These patterns provide solutions for the problems faced in the design of object-oriented applications. These solutions are described on a very abstract level and can be modified according to the needs of the design.

Exceptional success of software design patterns in object-oriented software development compelled software engineers to exploit the same for the design of parallel applications. For this book, we follow the design patterns for parallel programs explained by Mattson et al. They categorized the parallel design patterns into four design spaces, namely: Finding Concurrency, Algorithm Structure, Supporting Structures, and Implementation Mechanisms. The parallel patterns in the Finding Concurrency design space help designers divide the problem being addressed into many subproblems that can be solved in parallel. The patterns from the Algorithm Structure design space help mapping the subproblems defined in the Finding Concurrency design space onto a parallel algorithm. This algorithm is translated into a workable program coding scheme using the patterns from the Supporting Structure design space. In the end, the Implementation Mechanisms design space deals with process management and synchronization among the parallel tasks at the lowest programming level.

The use of design pattern facilitates the maintenance of large and complex software. However, complex applications may lack up-to-date documentation. For this purpose, researchers have developed techniques to identify design patterns from the structure of the application itself. It helps in making the application understandable and manageable. Design pattern identification in the UML diagrams of object-oriented applications is an active research field. Similarly, many state-of-the-art approaches exist to identify parallel patterns in both already parallelized applications and sequential applications. Parallel-pattern identification in already parallelized applications helps in improving performance with efficient thread mapping and optimized runtime environment.

Identification of parallel pattern identification in sequential applications helps easing the parallelization process. We discussed many approaches that identify parallel patterns and parallelize the sequential applications. However, some of these approaches rely only on static analysis which can limit the pattern identification due to the constraints of static analysis. Most of the approaches only identify one or two of the parallel patterns and some of them require specialized hardware to run the parallelized version efficiently. Our pattern identification framework uses a hybrid analysis to overcome the problems of both static and dynamic analysis. We identify almost all of the parallel patterns available in the algorithm structure design space and target general purpose applications running on commonly available multi-core hardware.

# 3  DiscoPoP: A Parallelism Discovery Tool

With the amount of parallelism now available on single-chip processors, the development of software capable of using these resources is becoming more important. The problem is to parallelize the complex legacy code produced before the parallel hardware era. One solution is to redesign the whole sequential application from scratch, but it can be a very time-consuming and sometimes never-ending task due to the lack of proper documentation being available. For this purpose, major research efforts are being made to develop tools for parallelism discovery and auto-parallelization, but they somehow fall short of the optimal outcome as we have discussed in Sections 1.4 and 2.5.

A major problem in parallelizing a sequential application is to fully understand its algorithm and code logic. It includes analysis of different language constructs used to implement the application and dependences among them. We have developed a parallelism discovery framework called DiscoPoP (Discovery of Potential Parallelism) [2, 9, 10]. DiscoPoP dissects the input code to a very basic level of program instructions and analyzes the majority of existing dependences. Based on the available information, it proposes different parallelization opportunities found in the input code.

An overview of the workflow of DiscoPoP has been provided in Figure 1.1. The dependence and computational-unit analyses are performed during the first phase of DiscoPoP . The dependence analysis includes both data and control flow analyses of the input code, while the computational-unit analysis identifies computational units in the input code. This chapter explains both of these analyses in detail. Li [2] provides the complete picture of this phase of DiscoPoP. We only provide a summary of these analyses here. We also present the framework for parallel pattern identification.

## 3.1  Dependence Analysis

A dependence between two instructions or code sections of a program exists when: both of them read from or write to the same memory location creating a data dependence; execution of one of them is dependent on the outcome of the other creating a control dependence. The biggest hurdle to overcome the parallelization of a sequential program is to resolve the dependences preventing parallelism. For this reason, dependence analysis is the basic component of a parallelization tool.

The dependence analysis is one of the two major parts of the first phase of the DiscoPoP framework. A profiler has been developed to provide the comprehensive information on both types of dependences in the input program. It employs a hybrid (both static and dynamic) approach to gather dependences. The DiscoPoP profiler is based on a compiler project called LLVM (Low Level Virtual Machine) [26]. LLVM provides a collection of tools for source-to-source translation and different types of code analysis. We explain the basics of LLVM in the following.

### 3.1.1 Introduction to LLVM

The LLVM project [26] was started at the University of Illinois as a project to develop a modern compiler. It would support many programming languages using both static and dynamic compilation capabilities. The project has grown over the years to become a collection of modular and reusable compiler and tool-chain technologies. At the start of the project, LLVM stood as the acronym for the project name Low Level Virtual Machine. However, the way the project progressed has very little to do with virtual machines. So now LLVM is not taken as an acronym but the full name of the project itself.

LLVM has become an umbrella project for many sub projects that are available for use. A compiler for a new language can be easily constructed using LLVM. Following are some of the important sub-projects of LLVM:

- The **LLVM core libraries** provide code optimizations and generators for many different architectures. These libraries are independent of the source and target language and other compilers can easily adapt to use these libraries to take advantage of available optimizations.

- The **Clang** project provides a C language family (C/C++/Objective-C) front-end for the LLVM project. It is considerably fast and aims to provide a platform for easier source-level tool development.

- The **compiler-rt** project consists of highly tuned routines that support low-level code generators. This project provides generators for the operations that are not directly supported by the target platforms.

- The **OpenMP** project provides the OpenMP implementation for Clang. The **polly** project provides a suite of auto-parallelism, vectorization and cache-locality optimizations using the polyhedral model.

- The **lld** project aims to provide a built-in linker for Clang and LLVM.

Some other popular sub-projects of LLVM are **libc++, libclc, klee, SAFECode** and **LLDB**. All of them add different functionalities to the LLVM framework. In addition to these official sub-projects, many individual projects are also available that provide support for many other languages like Ruby, Python and Java etc. LLVM is being used from compiling light-weight applications for embedded systems, to compiling Fortran codes running on big supercomputers.

LLVM has introduced a special representation of code that makes it independent of source and target languages. It is called *LLVM Internal Representation* (*LLVM IR*). Almost any programming language can be ported to LLVM by just implementing a front-end that translates the source language to IR. All of the LLVM provided optimizations and generators work around IR.

Listing 3.1: A sample code

```
1  int main(){
2      int a, b, c;
3      a = 10;
4      b = 20;
5      c = a + b;
6      return c;
7  }
```

Listing 3.2: LLVM IR representation of the sample code from Listing 3.1

```llvm
1 ; ModuleID = 'sample.c'
2 target datalayout = "e-m:e-i64:64-f80:128-n8:16:32:64-S128"
3 target triple = "x86_64-unknown-linux-gnu"
4
5 ; Function Attrs: nounwind uwtable
6 define i32 @main() #0 {
7 entry:
8   %retval = alloca i32, align 4
9   %a = alloca i32, align 4
10   %b = alloca i32, align 4
11   %c = alloca i32, align 4
12   store i32 0, i32* %retval
13   store i32 10, i32* %a, align 4
14   store i32 20, i32* %b, align 4
15   %0 = load i32* %a, align 4
16   %1 = load i32* %b, align 4
17   %add = add nsw i32 %0, %1
18   store i32 %add, i32* %c, align 4
19   %2 = load i32* %c, align 4
20   ret i32 %2
21 }
22
23 attributes #0 = { nounwind uwtable "less-precise-fpmad"="false"
        "no-frame-pointer-elim"="true" "no-frame-pointer-elim-non-leaf"
        "no-infs-fp-math"="false" "no-nans-fp-math"="false"
        "stack-protector-buffer-size"="8" "unsafe-fp-math"="false"
        "use-soft-float"="false" }
24
25 !llvm.ident = !{!0}
26
27 !0 = !{!"clang version 3.6.1 (tags/RELEASE_361/final)"}
```

LLVM IR uses *static single assignment* (SSA) form [99]. In the SSA form, each variable can be assigned only once in the whole program and every variable has to be defined before this assignment. So if a variable is assigned multiple times in the source code, it is given a new name at every new assignment in the SSA form. This makes application of many code optimizations, such as, *dead code elimination, constant propagation* and *value range propagation* etc. easier. LLVM IR uses an infinite number of registers and an IR instruction can have up to three addresses in it. This type of instruction set is called the *three-address code*.

The highest level structure of LLVM IR is called a *module* [100]. A module may consist of functions, global variables, and symbol table entries. The LLVM linker at the end of the compilation process merges different modules together to resolve forward declarations and missing symbol table entries. Listing 3.1 shows a simple function that declares two integers and adds their values into the third one. We used Clang to generate the LLVM IR for this code and the generated IR is shown in Listing 3.2.

The IR content starts with declaring a module, followed by the information about the target data layout and operating system. The actual body of the function is from Line 6 to Line 21. The keywords of the IR are very simple and straightforward. All the keywords starting with the % symbol are local variables (global variables start with the symbol @). The *alloca* instruction allocates a variable in memory with the provided data type. The *store* instruction stores the value into a variable. Similarly, the *load* instruction retrieves the value of a variable from memory. At the end, the IR content contains some information used by LLVM internally.

LLVM provides a *pass framework* used for developing passes to analyze and optimize the input code [55]. These passes can be derived from one of many sub-classes of the main *Pass* class. the following sub-classes are available in the pass framework:

- The *ImmutablePass* class is used for the passes that are never executed. Such passes usually provide the information about the current compiler's configuration and underlying target machines. They do not perform any transformations.

- The *ModulePass* class is used for the module level passes. The whole program is available for analysis and transformation in a module pass. These passes can add or remove functions from the program or perform program level analysis.

- The *CallGraphSCCPass* class is used by passes that traverse the programs bottom-up. These passes performs analysis on the strongly connected components (SCC) of a module. SCC is a well known concept in the graph theory [101]. They may change the functions available in the current SCC only.

- The *FunctionPass* class is used by passes to optimize all the functions in a program. A function pass is run on every function individually. All the information of the current function is available and other functions cannot be modified.

- The *LoopPass* class is used when a pass is required to run on every loop in the program. A loop pass can only get access to the current loop. All other loops are not accessible.

- The *RegionPass* class is used by passes for the optimization of a region. A region in LLVM is defined as a single-entry / single-exit code segment in a function. A region in the program may have other regions in it. A region pass is applied to the innermost region of the program first and then the region it is contained in. The outermost region is processed last.

- The *BasicBlockPass* class is used when the scope of a pass is limited to a basic block. LLVM defines a basic block as a code section with a single-entry / single-exit and without any conditional statements in it. A basic block pass can perform optimizations on the basic-block level.

- The *MachineFunctionPass* class is used by passes that generate the machine-dependent code. Their scope is at the function level and all the restrictions of a function pass applies to these passes, too.

LLVM provides a comprehensive library of different passes [102]. All the passes are generally classified into one of the two main categories. An *analysis* pass can generate reports of some types of analysis performed on the code. However, it does not change the code. On the other hand, a *transform* pass performs the optimizations and transforms the code accordingly.

All of the passes include a list of passes on which they depend and another list of passes whose results they invalidate. All of this information is crucial for LLVM's *pass manager*. This pass manager is responsible for correct scheduling of all the passes during the analysis and optimization of the compilation process.

## 3.1.2 Dependence Detection in DiscoPoP

The dependence analysis of DiscoPoP uses the pass framework of LLVM [2]. A new DiscoPoP pass is applied to the input code for this purpose during the compilation phase. This pass transforms the input code and inserts profiling instructions into the code. All the memory accesses, conditional statements (e.g., if and switch statement), and loops are instrumented. The profiling instructions call the appropriate functions defined in DiscoPoP's profiling library. These functions record the events happening at the runtime of the input application, for example, read or write of a memory location, iteration count of a loop, and end of a loop.

Listing 3.3: A sample code with nested loops

```
10 static int s[NUM] = {2, 5, 9, 4, 7, 1, 3, 6};
11 double r[NUM] = {0.0};
12
13 void main(void)
14 {
15     int i, j, count;
16     double k, t;
17     bool flag;
18
19     count = 0;
20     for (i = 0; i < NUM; i++)
21     {
22         count++;
23         flag = (r[i] >= 0.375);
24         for (j = 0; j < NUM; j++)
25         {
26             k = s[i] + s[j] * 0.25;
27             t = s[j] * 1.17;
28             r[j] += (k + t) * ((rand() % 10 + 1) / 10);
29         }
30     }
31 }
```

The DiscoPoP profiler is capable of reporting all the three types of data dependences, namely: RAW, WAR, and WAW. This is done by using signature-based profiling. The signature used in profiling is an array of fixed length. A hashing function is used to map memory addresses to the array indices. The profiling algorithm is straight forward. Two different signatures are maintained; one for read operations and the other for write operations. The following steps are taken to report dependences for a memory location $x$:

- A RAW dependence is reported if $x$ is read and there exists a record of writing to $x$ in the write signature.

- A WAW dependence is reported if $x$ is written and there exists a record in the write signature for $x$.

- A WAR dependence is reported if $x$ is written and there exists a record of reading from $x$ in the read signature.

The read signature for $x$ is also updated to the current source line in the event of a read operation. Similarly, write signature of $x$ is updated in case of a write operation with the current source line. This recording of the source line numbers enables the DiscoPoP profiler to enlist them in the final dependence-analysis report.

Listing 3.4: Dependences reported by the DiscoPoP profiler for the code in Listing 3.3

```
1 START 1:15
2 1:19 NOM   {INIT *}
3 1:20 BGN loop
4 1:20 NOM   {RAW 1:20|i} {WAR 1:20|i} {INIT *}
5 1:22 NOM   {RAW 1:19|count} {RAW 1:22|count} {WAR 1:22|count}
6 1:23 NOM   {RAW 1:20|i} {RAW 1:28|r} {WAW 1:23|flag} {INIT *}
7 1:24 BGN loop
8 1:24 NOM   {RAW 1:24|j} {WAR 1:24|j} {INIT *}
9 1:26 NOM   {RAW 1:20|i} {RAW 1:24|j} {WAR 1:28|k} {INIT *}
10 1:27 NOM   {RAW 1:24|j} {WAR 1:28|t} {INIT *}
11 1:28 NOM   {RAW 1:24|j} {RAW 1:26|k} {RAW 1:27|t} {RAW 1:28|r} {WAR 1:28|r}
     {INIT *}
12 1:30 END loop 8
13 1:31 END loop 8
14 END
```

Listing 3.3 shows a sample code with two nested loops. The dependences reported for this code by the DiscoPoP profiler are shown in Listing 3.4. Data dependences are represented in <sink, type, source> format. Whereas a `sink` is denoted by a pair <fileID:lineID> and a source is represented as a triple in <fileID:lineID|variableName> format. More than one data dependences on the same sink are aggregated. The NOM keyword denotes absence of any control flow information in this line of the report. On the other hand, BGN and END keywords represent the beginning and ending of a control region, for example, a loop. The INIT keyword represents the initialization of a variable.

Figure 3.1 shows the architecture of the DiscoPoP profiler. It is a multi-threaded design, where the main thread executes the instrumented application and produces the data from instrumentation. This data is distributed among the different worker threads. Each worker thread handles events related to a range of unique memory addresses assigned to it. A worker thread records data dependences locally and the results of all the worker threads are later merged together.

## 3.2 Computational Units

Parallelism is mostly available either at instruction-level or in programming language structures (e.g., loops or functions) level. The instruction-level parallelism is exploited by modern compilers, for

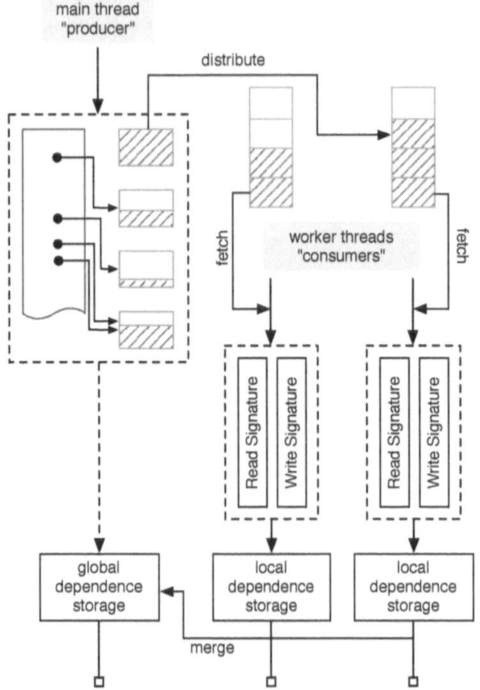



Figure 3.1: Architecture of the DiscoPoP data dependence profiler [2].

example Intel ICC compiler, efficiently [103]. Most of the approaches discussed in Chapters 1 and 2 detect parallelism in the different structures of different programming languages. However, the structures of two or more languages may not be similar. This limits the parallelism identification approaches using specific strucures.

To overcome this problem, DiscoPoP proposes *computational units* (CUs) [9, 58, 61, 66, 67, 104] as the basis for parallelism detection. In-length details of CUs are discussed in the doctoral book of Li [2]. For the sake of completeness, we only provide a definition, main characteristics, and the identification process of CUs here.

A CU follows the read-compute-write pattern. A set of instructions reads data from memory, performs computations over the read data, and the results are written back into memory at the end. A CU does not exhibit any meaningful parallelism in it, hence making it a good candidate to be the basis of our parallelism identification framework. The next section provides the formal definition of a CU.

## Definition

Let $C$ be a code section and $V_c$ be the set of variables that are global to $C$. Let $R_x$ be the set of instructions reading variable $x$ and $W_x$ be the set of instructions writing $x$. $C$ is a computational unit if:

$$\forall v \in V_c, R_v \to W_v \tag{3.1}$$

Where $\to$ is the happens-before relationship [105]. In simple words, a variable may not be read after it has been written by an instruction in the same CU. This property of a CU enables us to identify all the important flow and output dependences among global variables. These dependences are the most important ones to discover any parallelism opportunities in a program. On the other hand, as the granularity of a CU is small, all the dependences among any local variables in a CU does not exhibit any meaningful parallelism to be exploited by threads. Hence, these dependences are not reported to the user. The omission of these dependences may lose some data but the overhead of the parallelism discovery process is reduced considerably by this approach.

The writing of one variable after or before a read operation of another global variable does not invalidate a CU. parallel reads and writes of separate variables can be exploited by the instruction-level parallelism.

All the global variables read and written by a CU belong to its *read* and *write* sets, respectively. Both these sets can have common member variables. All the instructions of a CU reading variables in the read set makeup the read phase of the CU. Similarly, the write phase of a CU consists of all the instructions in the CU that are writing to the variables in its write set. All the instructions that are doing computations belong to the compute set of the CU. The identification of these phases simplify the mapping of the detected dependences from the code onto different CUs. The dependences among the instructions belonging to the read and write phases of a CU are ignored, as stated earlier.

## CU Identification

The CU identification in DiscoPoP is performed on the basic block level. A basic block in LLVM is defined as a code segment that does not contain any control branches. It means that if the control reaches the first instruction of a basic block, all the instructions inside it are guaranteed to be executed. This provides us the opportunity to identify the CUs at the finest level and extract most parallelism among them.

Algorithm 1 shows the pseudo-code of the implemented approach in DiscoPoP that follows the definition of a CU described in Statement 3.1. A new CU is created with the start of each basic block in the program. All the instructions in a basic block are traversed in the order as they occur in the program. The write instructions are recorded as write phase of the CU and variables written in these instructions are added to its write set. Additionally, if the used variable is global to the basic block, they are recorded in a set of suspicious variables.

On the other hand, if the current instruction is a read instruction, the variable being read is first checked in the suspicious variables set. If it is not an element of this set, the current instruction and variable are added to the read phase and read set of the CU, respectively.

However, If the variable read by the current instruction is an element of the suspicious variables set, it means that this variable has been written by an instruction that is a member of the write phase

```
cuVector = empty vector of CUs
suspiciousVars = empty vector of variables
for each basic block B in the program do
    globalVars = variables that are global to B
    cu = new CU
    cuVector.insert(cu)
    for each instruction I in B do
        if I is a write instruction then
            cu.addInstruction(I)
            v = variable written by I
            if v is in globalVars then
                suspiciousVars.insert(v)
                cu.writeSet+=v
                cu.writePhase+=I
            end
        end
        else if I is a read instruction then
            v = variable read in I
            if v is in suspiciousVars then
                // Violation of CU definition. Start a new CU.
                cu = new CU
                cuVector.insert(cu)
                suspiciousVars.clear()
            end
            cu.addInstruction(I)
            if v is in gloabalVars then
                cu.readSet+=v
                cu.readPhase+=I
            end
        end
        else
            cu.computeSet += I
            cu.addInstruction(I)
        end
    end
end
```

**Algorithm 1** : Algorithm to identify CUs.

```
1 int x = 3;
2 for (int i = 0; i < MAX_ITER; ++i) {
3       int a = x + rand() / x;
4       int b = x - rand() / x;
5       x = a + b;
6 }
```

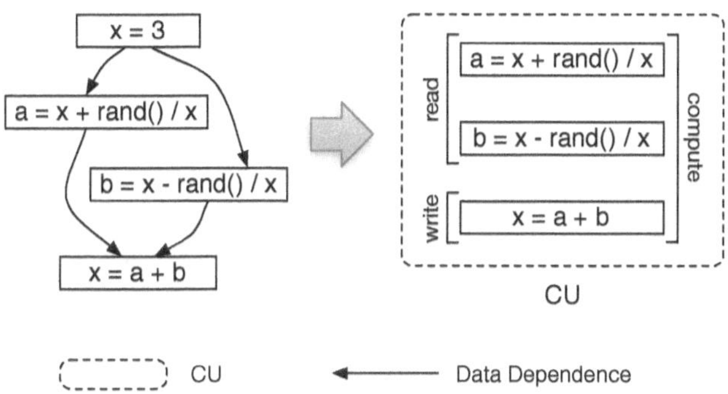


Figure 3.2: CU identification.

of the current CU. Hence, the current instruction cannot be added to this CU. In this case, a new CU is created and the set of the suspicious variables is set to null. Rest of the process remains the same. In the case, we encounter an instruction that is neither read nor write instruction, we add it to the compute set of the CU.

Figure 3.2 shows an example of the CU identification process. Lets consider the body of the loop starting at Line 3. Here, variable x is a global variable while variables a and b are both local variables. Hence the CU identification algorithm simply ignores both of them. The statements at Lines 3 and 4 are added to the same CU as they perform read operations on x, which has not been written by any statement belonging to the current CU yet. Note that the write operation on x in Line 1 belongs to another CU as it is outside the boundary of the current basic block. x is added to the read set of the CU, while Lines 3 and 4 are added to the read phase of the CU. Both of them are added to the compute set of the CU, too, as they are also performing some computations. The statement at Line 5 has a write operation on x that does not violate the CU definition. So it is also added to the current CU. x is added to the write set and Line 5 is added to the write and compute phase of the CU. The loop body ends at Line 6, so the current CU ends at Line 5. Any further code is assigned to new CUs accordingly.

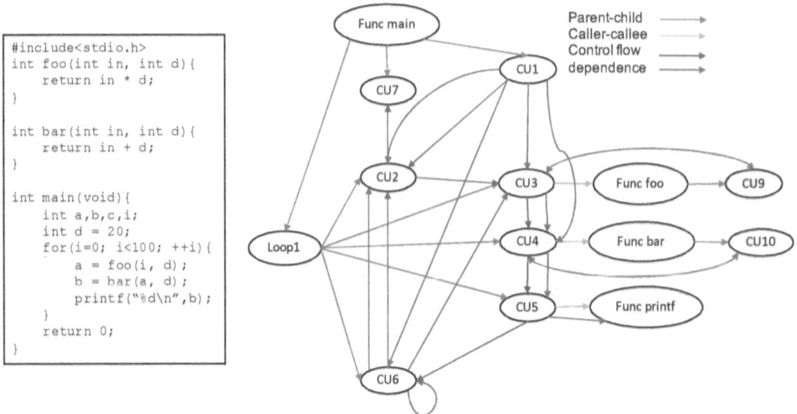



Figure 3.3: A simplified program execution graph.

## 3.3 Program Execution Graph

Different analyses performed by DiscoPoP produce different output, like CUs, control flow, and data dependences. All this information is important for parallelism discovery, however, we need to represent it in a meaningful way so that the parallelism discovery process becomes easier. For this purpose, we merge all the information produced by DiscoPoP into a single graph that we call *program execution graph* (PEG).

The vertices of a PEG belong to one of the three types: loop, function, or CU. A loop vertex represents a loop in the analyzed program. Similarly, a function in the program is represented by a function type vertex. All the CUs are represented by corresponding CU type vertices. All vertices may contain additional information, like actual source code lines of their lexical extent, size of the data read or written in a CU, number of LLVM-IR instructions, and number of iterations of a loop etc.

The edges of a PEG represent four types of relationships: parent-child, caller-callee, control flow and data dependence. The parent-child relationship shows all the children of a vertex that are in the lexical extent of the parent, for example CUs in the body of a loop. A function type vertex can be a parent of either loop or CU type vertices. A loop type vertex can be parent of other loop or CU type vertices. But, a CU cannot become a parent of any other vertex in the graph. To clarify, a recursive function call i handled by the caller-callee relationship and not the parent-child one.

The caller-callee relationship exists only between a CU type vertex as caller and a function type vertex as callee. It shows that the CU has a function call in it. The control flow edges represent the execution order among the child CUs of a function. The control flow analysis of the input program is done statically. Therefore, it is bound by the lexical extent of a function.

The data dependence type edges show existence of data dependences among the source and sink vertices. This relationship exists among CU type vertices only. Additional information like the size of the data transferred and variables involved in the dependences are also reported on their respective

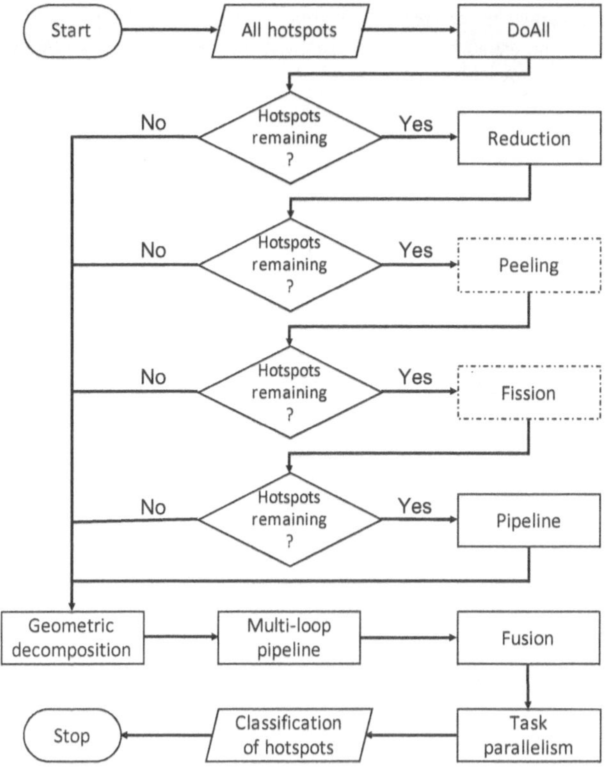



Figure 3.4: The order of pattern identification.

dependence edges. Figure 3.3 shows a simplified program execution graph of the code shown in the figure. All types of edges and vertices can be seen in the figure.

## 3.4 Framework for Parallel Pattern Identification

The complexity of a PEG even for a very small program, as shown in Figure 3.3, demonstrates how difficult it can be to understand the PEG of a large program. We require a comprehensive framework that can automatically identify the parallelism opportunities using a PEG as input. We motivate the need of such a framework in this section.

Our proposed framework tries to identify parallel patterns in hotspots in the order shown in Figure 3.4. A hotspot is a code section that takes most of the execution time of a program. The order of identification is designed to get the most beneficial parallelization opportunities reported first. Initially, the framework analyzes single loop patterns and optimizations. These include *doAll, reduction, peeling, fission,* and *pipeline*. Kennedy and Allan [24] has provided details of the optimizations named here. They are identified in the order specified and all the classified hotspots are removed from the unclassified hotspots list.

The identification of *peeling* and *fission* is considered to be a part of our future work. In the end, all the classified and unclassified hotspots are checked for the existence of *geometric decomposition, multi-loop pipeline, fusion,* and *task parallelism*. The identification of these three patterns in addition to the patterns already identified in the hotspots enables us to exploit nested parallelism.

Once the identification process is complete, our framework reports the pattern identification results. This report points out the locations in the source code where patterns have been identified successfully. Furthermore, the output also suggests the classification of CUs according to the skeleton of an appropriate supporting structure pattern. For example, for an identified pipeline, the output suggests which CUs should be grouped into the same pipeline stage. With the help of all this information, the sequential code can be easily transformed by a programmer to implement the identified parallel pattern.

The ultimate goal of our framework is to automatically transform a sequential application according to the identified patterns and get a parallelized version of the application. Techniques for an automatic transformation are reported in literature [98, 106]. Our future work includes a semi-automatic transformation of sequential applications into parallel ones based on the identified parallel patterns.

## 3.5 Summary

In this chapter, we introduced a parallelism discovery tool called DiscoPoP. DiscoPoP is based on LLVM, a compiler project. LLVM provides a pass framework that is based on different classes of passes. A pass derived from any of these classes can perform any specific type of program analysis. DiscoPoP uses this pass framework to gather the information required about the input program for parallelism identification.

DiscoPoP does dependence and control flow analysis. It also groups the statements of an input program into more manageable and meaningful units called computational units (CUs). A CU follows a read-compute-write pattern and does not exhibit any meaningful parallelism inside it. All the gathered information from the input program is later combined into a graph called program execution graph (PEG). This PEG of the input program is the basis of our parallelism identification process.

We motivated the need of a framework for identification of parallel patterns from the PEG of an input program and proposed an approach for such a framework. This proposed framework goes through all the hotspots of the program and tries to identify different patterns like geometric decomposition, do-all, and pipeline etc. At the end it reports the identified patterns and respective suggested locations in the source code.

# 4 Pattern Identification with Template Matching

Template matching is a widely used technique in the field of computer vision [107]. The goal of computer vision is the automatic detection of objects in digital videos and images. The defining features of an object being detected are transformed into a template vector and the input image is transformed into an image vector. template matching uses these two vectors to identify the occurrences of object in the image.

The template matching technique computes cross correlation [108] between a template vector $\vec{p}$ and a graph vector $\vec{g}$. Let $\vec{p}$ and $\vec{g}$ be of the same length, then the cross correlation is computed using the formula $CC = \vec{p} \cdot \vec{g}$ [72]. The value of $CC$ is maximized when both vectors have similar values, as their dot product amplifies the resultant value.

However, if the values of the elements of $\vec{g}$ are much larger than that of $\vec{p}$, then sometimes $CC$ of $\vec{p}$ and $\vec{g}$ can be bigger than the $CC$ of $\vec{p}$ and $\vec{p}$, which is an exact match. Thus the problem arises of finding the best match. This problem is solved by normalizing the value of $CC$. This is achieved by using the formula:

$$CC_n = \frac{\vec{p} \cdot \vec{g}}{\|\vec{p}\| \, \|\vec{g}\|} \qquad (4.1)$$

Geometrically speaking, the value of $CC_n$ represents the cosine function of the angel between vectors $\vec{p}$ and $\vec{g}$. When the vectors are parallel to each other, it implies that they are mostly similar to each other and the angel between them is $0°$. Therefore, $CC_n$ computes to $1$, the maximum value. On the other hand, if the vectors are perpendicular to each other, implying totally opposite to each other, the value of $CC_n$ is $0$. Any value of $CC_n$ between $0$ and $1$ shows the degree of similarity among the two vectors.

We build upon the template-matching technique introduced by Dong et al. [72]. There, both template and target program are represented by vectors and cross correlation between these two vectors is used to determine how similar they are. We adapt this concept to the identification of parallel patterns in program execution graphs (PEGs).

The templates used by Dong et al. are derived from UML diagrams of sequential patterns, which expose a fixed structure with a fixed number of components. On the other hand, the structure of parallel design patterns is also well defined, but the number of components in a pattern may vary significantly. For example, the number of stages in a pipeline may differ, depending on the nature of the application. Hence, our parallel pattern templates cannot be of fixed length as in the case of sequential patterns. A parallel pattern matches if certain conditions are met. For example, for a pipeline to exist there should be some separable stages inside a code section with repeating control flow such as a loop. Each stage should depend on the data produced by the previous stage. Obviously, such conditions are less rigid than those derived from fixed UML diagrams, which is why the size of our pattern templates must be adjustable. Also, template matching can only be used to identify such parallel patterns, where we can extract a good template vector for matching based on some definite characteristics of the pattern itself.

```
peg = getPEG(serialProgram)
Hotspots = findHotspots(peg)
for each h in Hotspots do
    g = getSubPEG(h)
    n = getNumberOfCUs(g)
    for each p in ParallelPatterns do
        p⃗ = getPatternVector(p, n)
        g⃗ = getPEGVector(p, g)
        CorrCoef[g, p] = CorrCoef(p⃗, g⃗)
    end
end
return CorrCoef
```

**Algorithm 2** : Parallel pattern identification using template matching.

Algorithm 2 shows the overall workflow of our approach for identifying patterns with template matching. We first look for hotspots in the input program—sections such as loops or functions that have to shoulder most of the workload. For each hotspot and pattern, we then create a pattern vector $\vec{p}$, whose length is equal to the number $n$ of CUs in the hotspot. The pattern vector plays the role of the template to be matched to the program. After that, we create the pattern-specific graph vector $\vec{g}$ of the hotspot's PEG sub-graph, which represents the part of the program to which the pattern vector is matched. Vectors $\vec{p}$ and $\vec{g}$ are derived from adjacency matrices reflecting dependences in the pattern and in the PEG, respectively. As a next step, we compute the value of $CC_n$ of the two vectors using Equation 4.1.

The $CC_n$ of the pattern vector and the graph vector of the selected section tells us whether the pattern exists in the selected section or not. As mentioned before, the value of $CC_n$ is always in the range of $[0, 1.0]$. A $1.0$ indicates that the pattern exists fully, whereas a $0$ indicates that it does not exist at all. A value in between shows the pattern can exist but with some limitations which we need to work around. Our tool points us to the dependences that cause the value of the correlation coefficient of a pattern to be less than $1.0$. This helps the programmer to resolve these specific dependences if he wants to implement that pattern.

Our tool not only indicates whether a parallel design pattern has been found in some section of the program but also shows how the code must be divided to fit the structure of the pattern. This helps the programmer implement the pattern and create a parallel version of the sequential input program. One should keep in mind that dynamic dependence profiling is never completely accurate and some dependences may not be reported due to the input sensitivity. Though we mitigate the input sensitivity by using different inputs, the programmer still needs to verify the correctness of the transformed parallel code. Template matching is used for the identification of pipeline and doAll patterns, which are discussed in the following.

## 4.1 Pipeline

The pipeline pattern is the most effective pattern to parallelize loops with inter-iteration dependences. Such loops are otherwise very difficult to parallelize. The idea is to divide the body of a loop into stages and assign them to different threads or processors as shown in Figure 2.5.

The performance of a pipeline is bound by the number of stages and load distribution among them. The pipeline pattern has been discussed in detail in Chapter 2. The definite structure of the pipeline pattern (i.e., stages with dependence between them) allows us to use these features for extracting a template vector and identify it in the hotspots by employing template matching. The following section explains our approach in detail.

### Identification Approach

Implementing a pipeline only makes sense if its stages are executed many times. For this reason, we restrict our pipeline identification to loops and recursions. In order to find a pipeline, we first let DiscoPoP deliver the PEGs of all hotspots. Because DiscoPoP counts the number of read and write instructions executed in each loop or function, we currently use this readily available metric as an approximation of the workload when searching for hotspots. A more comprehensive criterion, including execution times and workload, will be implemented in the future. We then compute an adjacency matrix for each hotspot graph, which we call the *graph matrix*. For each graph matrix, we create a corresponding pipeline pattern matrix of the same size, which we call the *pipeline matrix*. Pipeline matrices encode a very specific arrangement of dependences expected between CUs. For example, there must be a dependence chain running through all CUs in the graph because a pipeline consists of a chain of dependent stages. This specific property helps us to derive the pipeline pattern vector from the matrix.

Listing 4.1: Main loop of bodytrack.

```
1    for(i=0; i<num_of_frames; ++i} {
2        if{!pf.Update(i)) {
3            return 0;
4        }
5            pf.Estimate(estimate);
6            WritePose(output, estimate);
7            if(outputBMP){
8                outputBMP(estimate);
9            }
10   }
```

To make our approach easier to understand, we explain it using an example: the bodytrack application from the PARSEC 3.0 benchmark suite [109]. Bodytrack tracks a 3D pose of a marker-less human body with multiple cameras through an image sequence. Listing 4.1 shows the main loop of the bodytrack. Inside the loop body, line 2 retrieves an input frame, line 5 processes it and line 6 writes the output to a file. Optionally, line 8 writes a bmp image.

For this loop, DiscoPoP returns the CU graph shown in Figure 4.1(a). The loop body consists of three CUs: $CU_{update}$ covering lines 2 and 3, $CU_{estimate}$ covering line 5, and $CU_{output}$ covering lines

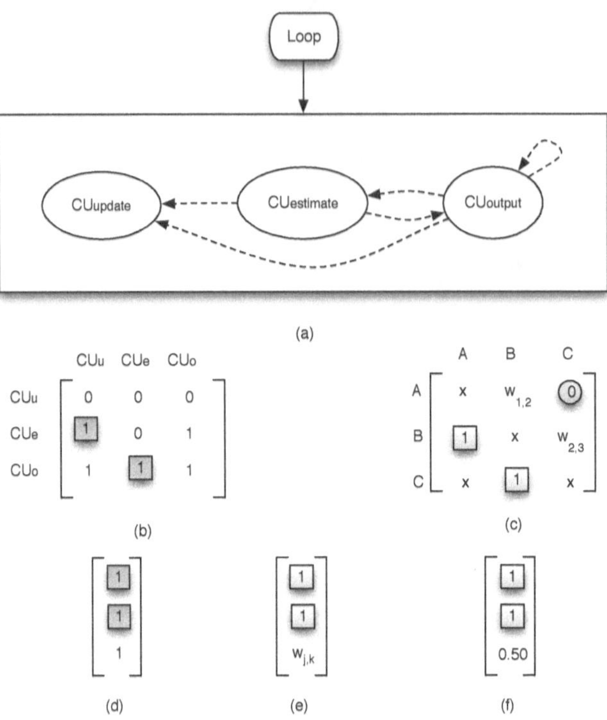

Figure 4.1: (a) PEG for Listing 4.1. (b) Corresponding graph matrix. (c) A 3x3 pipeline matrix. (d) Graph vector of graph matrix. (e) A 3-stage pipeline vector with weight variable $w_{j,k}$. (f) The resulting pipeline vector with $w_{j,k}$ value $0.50$.

6–8. The corresponding graph matrix of size 3x3 is shown in Figure 4.1(b).

According to the dimensions of the graph matrix, we create a pipeline pattern matrix of size 3x3, as depicted in Figure 4.1(c). We call it the *pipeline matrix*. Each row or column of this matrix represents a stage in the pipeline and the entries of the matrix represent dependences between them. Note that a pipeline matrix of any arbitrary number of stages can be created using the same pattern as shown in Figure 4.1(c). The entries of the pipeline matrix have the following specific meaning:

- An "x" means *don't care*, either 1 or 0 can be in its place. These dependences do not affect pipeline creation.

- A 1 indicates a mandatory dependence. The 1 entries together represent the chain of dependences along the stages of the pipeline and indicate the transfer of data among the stages. We call them the *chain dependences*.

- All the entries in the right upper triangle of the pipeline matrix, except the one in the last column of the first row, denote forward dependences. A *forward dependence* exists if a stage $S_j$ of the $i$th iteration of a pipeline depends on the result of a stage $S_k$ from its $i-1$th iteration, where $j < k$. This means that a forward dependence adversely affects the execution of the pipeline because an earlier stage of an iteration has to wait for the results of a later stage of the previous iteration. The $w_{j,k}$ in the pipeline matrix indicates forward dependences in the pipeline. Each individual $w_{j,k}$ quantifies the weight of this forward dependence. We calculate the value of each weight $w_{j,k}$ using the following formula:

$$w_{j,k} = 1 - \frac{(k - j)}{(\#stages - 1)} \tag{4.2}$$

  $\#stages$ represents the total number of stages in the pipeline. The weight decreases as the distance between two stages with forward dependences increases. Increasing distance between two stages with a forward dependence implies lower pipeline performance because the next iteration has to wait longer for the result of a later stage of the previous iteration.

- The 0 in the last column of the first row (encircled) of the pipeline matrix ensures that the first stage does not depend on the last stage because such a dependence would effectively serialize the pipeline. To rule out a serial pipeline, we always first verify that the graph matrix also contains the mandatory 0 in the last column of the first row.

The next step is to create the *graph vector* and the *pipeline vector* from the graph matrix and the pipeline matrix, respectively, and to calculate their correlation coefficient. To create these vectors we use the chain and forward dependences. Figure 4.1(d) shows the graph vector and Figure 4.1(e) shows the pipeline vector. These vectors are created as follows.

We fill the vectors from top to bottom with the exception of the last element using the entries in chain dependence locations of their corresponding matrices. In both cases, the last vector element is reserved for forward dependences. If there is no forward dependence in the graph matrix, the last element of both vectors becomes 0. If there is at least one forward dependence present in the graph matrix, the last element of the graph vector becomes 1 and the last element of the pipeline vector becomes the minimum across all forward-dependence weights in the pipeline matrix whose corresponding entry in the graph matrix is non-zero.

Figure 4.1(f) shows the pipeline vector after calculating all weights and choosing the minimum or zero. In general, the smaller the weight value in the pipeline vector the smaller is the value of the correlation coefficient. Finally, we calculate the correlation coefficient between the two vectors according to Equation 4.1, which tells us whether a pipeline exists in Listing 4.1 or not. We identified a three-stage pipeline pattern in the loop. The three stages correspond to the three CUs in the CU graph in Figure 4.1(a).

In our example, the correlation coefficient is $0.96$. The value of the correlation coefficient is less than $1.0$, showing that the identified pipeline has a forward dependence which needs to be resolved. Therefore, some additional code restructuring is needed to implement the pipeline. One way to resolve a forward dependence in a pipeline is to merge all the stages in the loop created by the forward dependence. Another solution is to add synchronization between the stages which are affected by the forward dependence.

The diagonal entries of a graph matrix represent self dependences of CUs. A self dependence of a CU in a loop is a loop-carried dependence. If a pipeline stage does not have any loop-carried dependence, then multiple instances of this stage can run in parallel. Such a stage is called a parallel stage. Our graph matrix of the hotspot for which a pipeline is identified readily shows which stages are parallel stages by taking their diagonal entries into account.

Our output of the identified pipeline pattern includes the necessary information about the division of code blocks into pipeline stages. Listing 4.2 shows the simplified output of the pipeline identification for the main loop of bodytrack shown in Listing 4.1 on Page 65. We deliver our results in the well-known extensible markup language (XML) format [110]. This makes the processing of our output by any other tool for other purposes a simple task. The first line of Listing 4.2 shows the relevant information of the loop being reported, followed by the details of the identified pattern. For the pipeline pattern, all the identified stages are listed. Each stage has a list of source code line numbers assigned to it. The dependences for each pipeline stage are also reported.

Listing 4.2: Simplified pipeline identification results for the main loop of bodytrack.

```
1    <loop id="34" start="32:350" end="32:360"  workload="88.62%">
2        <pattern name="pipeline">
3            <StageList>
4                <Stage id="_0">
5                    <Lines>32:351, ...</Lines>
6                    <Deps></Deps>
7                </Stage>
8                <Stage id="_1">
9                    <Lines>32:356, ...</Lines>
10                   <Deps>_0, _2</Deps>
11               </Stage>
12               <Stage id="_2">
13                   <Lines>32:357, ...</Lines>
14                   <Deps>_0, _1, _2</Deps>
15               </Stage>
16           </StageList>
17           <CC>0.96</CC>
18       </pattern>
19   </loop>
```

### Recursive function pipeline

A pipeline may also be created from a recursive function. The only precondition is that the current recursive invocation does not depend on the results generated by subsequent recursive invocations, as shown in the function shown in Figure 4.2(a). The function in Figure 4.2(b) calls itself and then uses the results generated by those recursive calls in further computations. The usage of recursion results prevents a pipeline because we need the next recursive invocation to complete before the current invocation can continue. To implement an identified pipeline for a recursive function, the code has to be restructured into a loop, which requires additional code refactoring. In the evaluation section, we show how we identify a pipeline in a recursive function of one of our test programs.

```
void foo(int a){
  if(a>0)
    foo(a-1);

  bar(a);
}
```

```
int foo(int a){
  int b = 0;
  if(a>0)
    b=foo(a-1);

  return bar(b);
}
```

(a)                                          (b)

Figure 4.2: Recursive functions: (a) pipeline can be implemented; (b) pipeline cannot be implemented.

## 4.2 DoAll

The doAll parallel pattern is a specific case of the task parallelism pattern. It exists in a loop that does not have a loop-carried or inter-iteration dependences. The parallel version of such a loop can execute all loop iterations in parallel to each other.

The performance of doAll mostly scales with the availability of parallelism in the hardware. Issues like false sharing, load imbalance, and imperfect scheduling can hinder the optimal parallelization using doAll. From a programmer's perspective, the main issue is to identify whether there exists a loop-carried dependence in a target loop. We identify loops with doAll pattern using template matching.

### Identification Approach

A loop can be parallelized according to the doAll pattern if there are no loop-carried or inter-iteration dependences. A forward or self-dependence is always loop-carried, as the control flow within a loop iteration moves in forward direction, which is why dependences within the same iteration must point backward. Note that inner loops in loop nests, which may reverse the control flow direction whenever a new inner iteration starts, are treated separately. The absence of forward or self-dependences is easy to verify based on the graph matrix, whose upper triangle shows all forward and self-edges (encircled in Figure 4.3).

Of course, there may also exist loop-carried dependences in backward edges of the graph matrix. However, to reliably distinguish them from intra-iteration dependences, our dependence profiler would have to record the iteration number along with each memory access, substantially increasing its memory overhead. On the other hand, loop-carried dependences in backward direction that are not accompanied by dependences in forward direction in the same loop are very rare. Basically, the absence of forward and self-dependences in a loop is a good indicator of the absence of loop-carried dependences. For this reason, we decided to refrain from the costly classification of backward dependences into loop-carried or not loop-carried.

The pattern vector $\vec{p}$ for a doAll pattern is of length equal to the number of elements of the upper

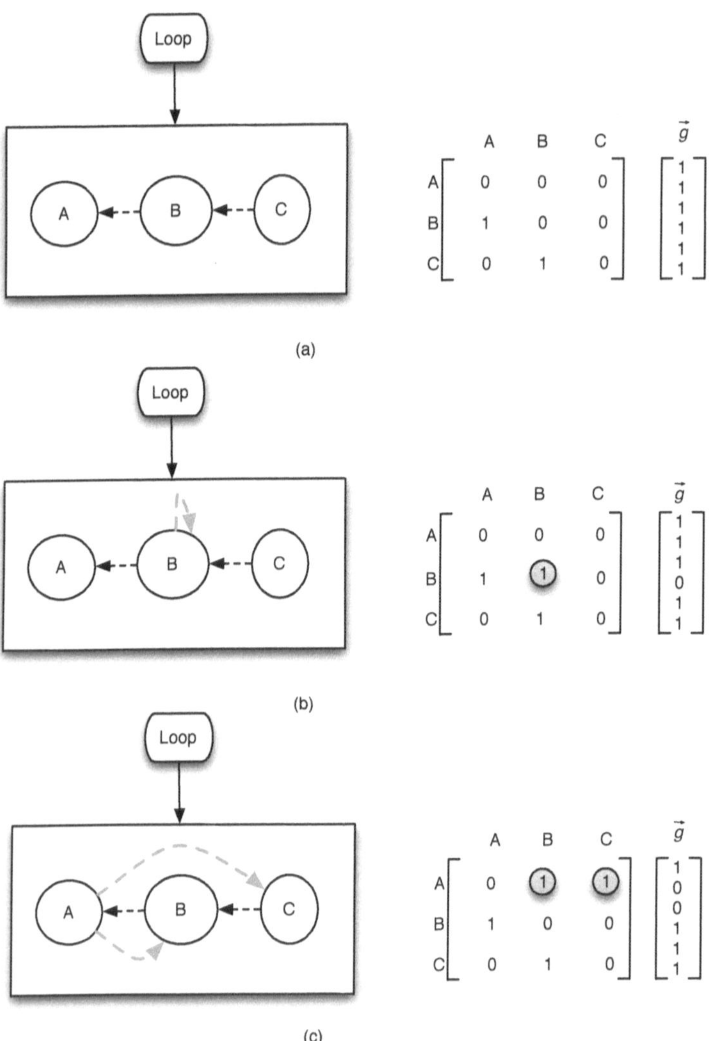

Figure 4.3: Example of loops for doAll identification: (a) doAll loop; (b) no doAll loop due to self-edge; (c) no doAll loop due to forward edge.

Table 4.1: Hotspots in test cases.

| Application | Benchmark Suite | LOC | Hotspots | Exec inst%* |
|---|---|---|---|---|
| bodytrack | PARSEC | 7698 | 7 | 88.62% |
| blackscholes | PARSEC | 914 | 2 | 99.75% |
| dedup | PARSEC | 3347 | 2 | 99.88% |
| ferret | PARSEC | 11263 | 2 | 99.96% |
| LibVorbis | NA | 54743 | 2 | 99.80% |
| c-ray | Starbench | 461 | 1 | 99.99% |
| md5 | Starbench | 407 | 1 | 84.15% |
| ray-rot | Starbench | 831 | 2 | 99.38% |
| rgbyuv | Starbench | 269 | 1 | 99.99% |
| rotate | Starbench | 448 | 1 | 98.31% |
| tinyjpeg | Starbench | 1241 | 1 | 99.99% |
| alignment | BOTS | 612 | 2 | 99.99% |
| nqueens | BOTS | 82 | 1 | 100.00% |

* Percentage of executed read and write instructions covered by the identified hotspots.

triangle of the graph matrix. All elements in $\vec{p}$ are initialized with 1. All elements of the upper triangle of the graph matrix are taken as elements of the graph vector $\vec{g}$, with their values being reversed from 0 to 1 and vice versa. Figure 4.3 shows some loop graphs along with their graph vectors $\vec{g}$.

Once we have determined the graph vector of our target loop, we calculate the correlation coefficient according to Equation 4.1. If the correlation coefficient is 1.0, the loop is a doAll loop. A correlation coefficient of less than 1.0 means that the loop cannot be parallelized with the doAll pattern in its current form due to some loop-carried dependences. The closer the value of the correlation coefficient is to 1.0, the fewer the number of loop-carried dependences. These loop-carried dependences are highlighted in the output of the tool and the programmer can try to resolve them manually, if the parallel implementation using doAll pattern seems profitable.

## 4.3 Evaluation

We tested our approach on three benchmark suites, namely: PARSEC 3.0 benchmark suite [109], Starbench [111], and the Barcelona OpenMP Task Suite (BOTS) [112]. We also evaluated with the open-source audio encoder LibVorbis [113]. All of the benchmarks in the three benchmark suites are shipped with a sequential and a parallel version. We report here all the benchmarks where our approach identified one or more pipelines or doAll patterns. Our evaluation did not report any false positive or false negative cases. One reason can be the smaller size of the benchmarks tested.

The time and memory consumption of the pattern identification heavily depends on the depth of the program execution graph. On average, the profiling with DiscoPoP causes a slowdown of

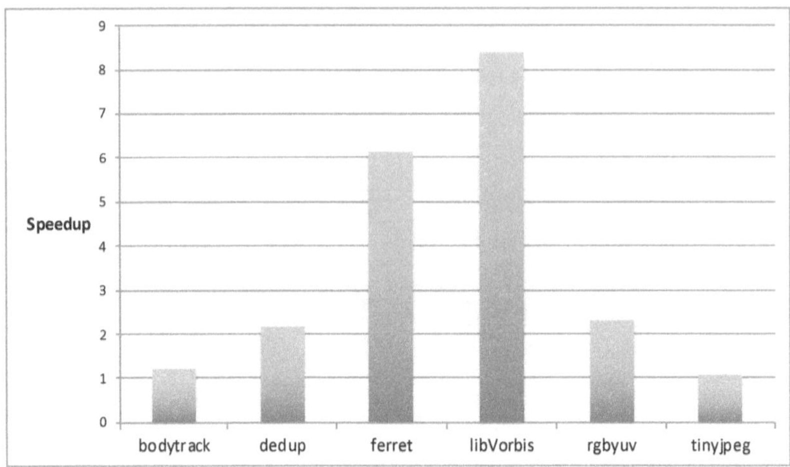

Figure 4.4: Average speedups of pipeline implementations in all applications.

137x [2]. Our pattern identification process is done offline and in comparison to the original runtime of the benchmarks showed an average of 62x higher runtime. The pattern identification consumed 152.95 MB of memory on average for all the six applications compared with 540 MB consumed by the profiler. The pattern identification was applied to all hotspots (sections bearing most of the workload) in these six applications. Table 4.1 summarizes the hotspots found in our test cases.

### 4.3.1 Pipeline

We tested all of the candidate programs for the applicability of pipeline patterns. We implemented the identified pipelines in all the programs. We also compared the results obtained for bodytrack, dedup, ferret and tinyjpeg with their pipeline versions, which are already available. The speedups of the implemented pipeline versions were measured on an Intel Xeon X5670 CPU with six cores and hyper-threading enabled. There were no pipelines identified in blackscholes. Table 4.2 gives an overview of the identified pipelines along with their correlation coefficients, while Figure 4.4 shows the average speedups achieved after implementing the identified pipelines. All the speedups were measured using 12 threads.

**Bodytrack**

In one of bodytrack's hotspots, we identified a pipeline pattern with a correlation coefficient of 0.96. Additional restructuring of the code was needed. It is the loop shown in Listing 4.1 on Page 65, which iterates over all input frames to track the body of a person. It covers 88.62% of the executed read and write instructions. As shown in Figure 4.1, the pipeline consists of three stages. $CU_{update}$ updates a frame, $CU_{estimate}$ calculates estimates and $CU_{output}$ writes the data to a file. Due to the

Table 4.2: Identified pipelines with correlation coefficients.

| Application | Pipelines identified | Corr coef | Pipelines implemented | Pipelines in Suite |
|---|---|---|---|---|
| bodytrack | 1 | p1: 0.96 | 1 | 1 |
| blackscholes | 0 | NA | 0 | 0 |
| dedup | 2 | p1: 1.00<br>p2: 1.00 | 2 | 1 |
| ferret | 2 | p1: 1.00<br>p2: 1.00 | 2 | 1 |
| LibVorbis | 2 | p1: 1.00<br>p2: 1.00 | 1 | NA |
| rgbyuv | 1 | p1: 1.00 | 1 | 0 |
| tinyjpeg | 1 | p1: 1.00 | 1 | 1 |

forward dependence between $CU_{estimate}$ and $CU_{output}$, we analyzed the code and decided to merge these two CUs into a single stage, forming a two-stage pipeline. We achieved an average speedup of 1.2 for this implementation because both stages of the pipeline were serial stages.

The parallel version of bodytrack available in PARSEC has indeed been implemented with two stages: an update stage which reads the data, and an estimate & output stage. This example demonstrates that a forward dependence does not necessarily present an insurmountable obstacle to the formation of a pipeline. Sometimes, a combination of merging stages and shifting the responsible instructions into other stages does the trick.

Listing 4.3: Main hotspot loop of dedup.

```
1    while{!EOF}{
2        readInput(Buffer);
3        Fragment(Buffer);
4
5        while{Blocks = Split(Buffer)}{
6            Refine(Block);
7            Deduplicate(Block);
8            Compress(Block);
9            Reorder(Block);
10       }
11   }
```

**Dedup**

Dedup is a program for data compression with data deduplication. Pipeline patterns were reported in two hotspots of dedup. These hotspots are the two nested loops shown in Listing 4.3. They cover almost all of the executed read and write instructions. In the outer loop, we found a three-stage

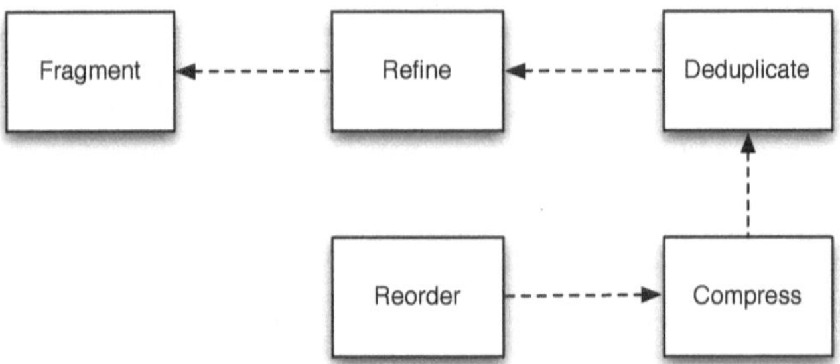



Figure 4.5: Five-stage pipeline in the parallel version of dedup.

pipeline with `readInput()` and `Fragment()` as the first two stages and the entire inner loop as the last stage. The correlation coefficient of this pipeline is 1.0. We identified another pipeline, this time with four stages, in the inner loop: `Refine()`, `Deduplicate()`, `Compress()` and `Reorder()`. The correlation coefficient of the second pipeline is also 1.0. We implemented both pipelines nested into one another. An average speedup of 2.3 was achieved. All of the stages identified were serial stages.

The existing parallel version of dedup in PARSEC runs a flat five-stage pipeline, which is shown in Figure 4.5. It covers the entire loop nest from Listing 4.3 except for reading the input at the entry of the outer loop. Instead of reading it periodically, the parallel version reads the input in one piece at program start. The remaining stages we identified in dedup match those of the benchmark's parallel version exactly.

The speedups of our own nested-pipeline version and of the flat-pipeline version from the benchmark were almost the same. The reason is that most of the work is done by the inner loop with the loop-carried dependence, which makes it a bottleneck. So the decision of creating a nested vs. a flat pipeline does not affect the speedup.

**Ferret**

Ferret traverses an image database and reports all images that appear similar to an input image. The serial version of ferret searches recursively through all sub-directories of an input directory. Listing 4.4 shows the workflow of ferret. Functions `scan_dir()` and `dir_helper()` traverse the directory tree, while `do_query()` compares images and writes its findings to a file.

We found two hotspots in ferret. `scan_dir()`, the first one, checks the entries of a directory and calls `do_query()` if an image was found or calls `dir_helper()` to proceed in child directories. We found a two-stage pipeline in this function, which is depicted in Figure 4.6(a). The correlation coefficient of this pipeline is 1.0. After carefully examining the code, we decided that `dir_helper()` should be moved to stage 1. This is because it calls `scan_dir()` again, which marks the start of the next pipeline iteration.

`do_query()`, the second hotspot, is nested inside the first one and covers 96% of the total

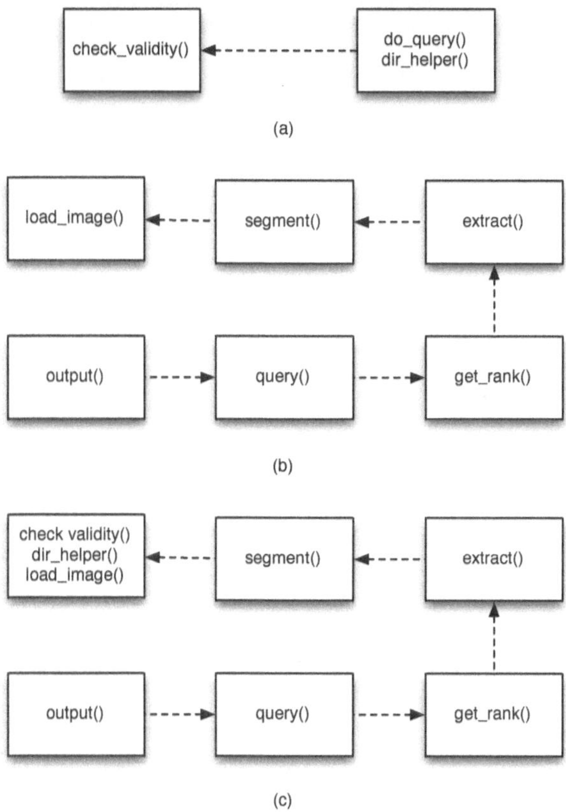



Figure 4.6: Pipeline in ferret: (a) pipeline identified in `scan_dir()`; (b) pipeline identified in `do_query()`; (c) pipeline implemented in PARSEC.

read and write instructions. There, we found a six-stage pipeline with a correlation coefficient of 1.0, which is illustrated in Figure 4.6(b). We implemented the nested pipelines by refactoring the recursion into iterations and achieved a speedup of 6.14.

We compared the reported pipelines with the parallel implementation in PARSEC. The PARSEC version features only one pipeline, which is shown in Figure 4.6(c). It has six stages and is essentially a combination of the two we identified, similar to our findings for dedup. The first stage of the benchmark's implemented pipeline merges the entry stages of the two identified pipelines. This is advisable because their workload is so small. The remaining implemented stages correspond to the last five stages of the second pipeline we found.

Listing 4.4: Pseudo code of ferret.

```
1    scan_dir( ent, path){
2        check_validity(ent, path);
3        if(IS_FILE(name)){
4            do_query(name);
5        }
6        else if(IS_DIR(name)){
7            dir_helper(ent, path);
8        }
9    }
10
11   dir_helper( ent, path){
12       dir = get_list(path+ent);
13       while (1) {
14           file = dir.get_next();
15           if (file) {
16               scan_dir(file.name, dir);
17           }
18           else{
19               break;
20           }
21       }
22   }
23
24   do_query(name){
25       image = load_image(name);
26       mask = segment(image);
27       dataset = extract(segment);
28       result = query(dataset);
29       rank = get_rank(result);
30       output(file, name, result, rank)
31   }
```

The difference between the speedups of the two implementations with single flat and nested pipelines, respectively, was negligible. The reason was same as in the case of dedup. This comparison demonstrates that our division into CUs is not always optimal. To improve the load balance between pipeline stages, future versions of our tool will merge the CUs of adjacent pipeline stages if appropriate.


**LibVorbis**

LibVorbis [113] is an audio encoding and decoding library, which converts wave files into compressed Ogg files. We used version 1.3.3 for our study. We analyzed a client application distributed alongside the library, which uses the library for conversion purposes. Figure 4.7 shows the encoding workflow of libVorbis. We found two hotspots in the form of two loops with one of them nested inside the

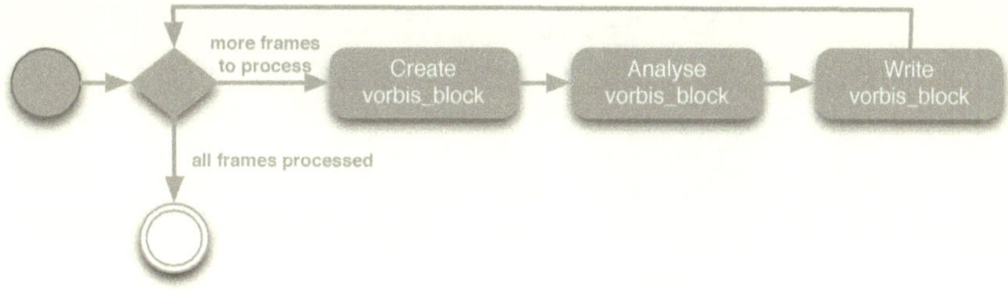

Figure 4.7: Workflow for encoding with libVorbis.

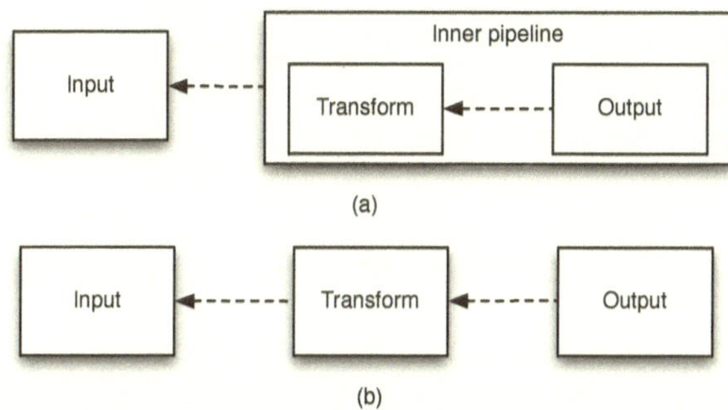


Figure 4.8: Implemented pipelines of LibVorbis: (a) nested and (b) merged.

other—very similar to the code in Listing 4.3. In the outer loop, we identified a pipeline with two stages: the first one for data input and the second one covering the entire inner loop. In the inner loop, we identified another pipeline with two stages, the first one for the transformation of the data and the second one for their output. The correlation coefficient of both pipelines was 1.0.

We decided to implement two parallel versions: one with nested pipelines and one with the two pipelines merged into a single flat three-stage pipeline, as shown in Figure 4.8(a) and (b), respectively. The Transform stage in both cases was identified as a parallel stage. Again the comparison of the speedups of the two implementations followed the results for dedup and ferret. The speedups of the two implementations were almost identical. The reason is that the Transform stage consumes most of the runtime in both cases. With three different input files, we achieved an average speedup of $8.4$. Figure 4.9 shows the speedups achieved for the different number of threads.

**rgbyuv**

This application converts an RGB image into the YUV format. The hotspot loop in the function `processImage()` in the rgbyuv benchmark is shown in Listing 4.5. We found a three stage pipeline

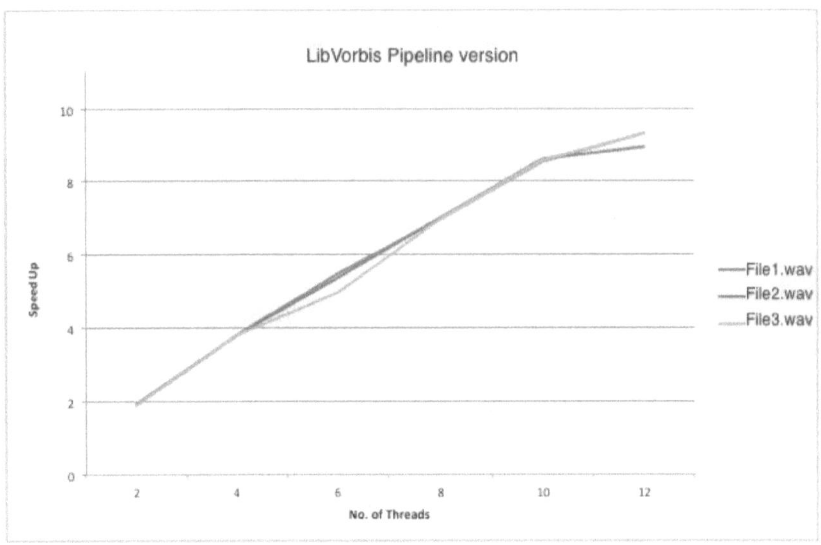

Figure 4.9: Speedup of parallel version of libVorbis.

in this loop. The first stage, consisting of lines 2, 3, and 4, and the last stage consisting of lines 10, 11, and 12, were serial stages and the middle stage, consisting of lines 6, 7, and 8 was classified as parallel stage.

Listing 4.5: Hotspot loop of *rgbyuv* benchmark.

```
1    for(int i = 0; i < args->pixels_to_process; i++) {
2        R = *in++;
3        G = *in++;
4        B = *in++;
5
6        Y = round(C1*R+C2*G+C3*B) + 16;
7        U = round(C4*R-C5*G+C6*B) + 128;
8        V = round(C7*R-C8*G-C9*B) + 128;
9
10       *pY++ = Y;
11       *pU++ = U;
12       *pV++ = V;
13   }
```

The inter-iteration dependences in the first and last serial stages are due to the increment of pointers. This loop can be converted easily into a do-all loop if array indexing is used instead of pointers. We examined the parallel version of *rgbyuv* in Starbench and found a do-all implementation of this loop. However our tool does not suggest resolving dependences. Instead it identifies the

Table 4.3: Identified doAll loops with a correlation coefficient of 1.0.

| Application | Do all identified | Do all implemented | Identified by Intel ICC |
|---|---|---|---|
| bodytrack | 6 | 6 | 0 |
| blackscholes | 2* | 1 | 2 |
| c-ray | 1 | 1 | 0 |
| md5 | 1 | 1 | 0 |
| ray-rot | 2 | 2 | 0 |
| rotate | 1 | 1 | 0 |
| alignment | 2 | 2 | 0 |
| nqueens | 1 | 1 | 0 |

* One doAll loop is not an intrinsic part of blackscholes.

patterns based on the current input code. The speedup for the identified pipeline was 2.29 compared to 10.29 of the do-all implementation. The difference between the speedups of both patterns is due to the way these patterns execute. This shows that the programming style of a programmer can have an impact on the identification of the parallel patterns.

**tinyjpeg**

The tinyjpeg benchmark decodes JPEG type images into the RGB format. We identified a two stage pipeline in tinyjpeg in a loop shown in Listing 4.6 in the function `convert_one_image()`. The first stage reads the input for conversion so it was a serial stage. The second stage converted the image to jpeg.

Listing 4.6: Hotspot loop of the *tinyjpeg* benchmark.

```
1   for(i=0; i<ntasks; i++) {
2       create_jdec_task(jpc, &jtask, i);
3       decode_jpeg_task(jdc, &jtask);
4   }
```

Both of the stages had inter-iteration dependences, so both were classified as serial stages. We achieved a speedup of 1.05 for the identified pipeline. We examined the parallel version of *tinyjpeg* available in Starbench and found a pipeline implemented for the same loop.

### 4.3.2 DoAll

Table 4.3 summarizes the the results of the doAll pattern identification, where the correlation coefficient was 1.0. We compared the identified doAll patterns with the parallel implementations in PARSEC and with the doAll loops identified statically by the Intel C compiler (ICC) [103] version 14.0.3. ICC is one of the sophisticated available compilers that can perform static doAll parallelization.

Table 4.4: Identified hotspot loops with a doAll correlation coefficient of less than $1.0$.

| Application | Hotspot loops | Corr coeff |
|---|---|---|
| bodytrack | 1 | d1: 0.98 |
| dedup | 2 | d1: 0.99<br>d2: 0.94 |
| LibVorbis | 2 | d1: 0.96<br>d2: 0.70 |

We did not find any profile-guided tool available to compare our results with. ICC found doAll loops only in blackscholes. We could not find any doAll loops in hotspots of dedup and LibVorbis with a correlation coefficient of $1.0$. Table 4.4 shows the hotspot loops whose correlation coefficients for doAll were less than $1.0$. The value of the doAll correlation coefficient depends on the number of loop-carried dependences and the total number of CUs in a loop. There were no doAll loops found in the ferret benchmark. The detailed results are presented below.

**Bodytrack**

We classified six loops in bodytrack as doAll loops, each with a correlation coefficients of $1.0$. All of these loops exist in functions called from the loop shown in Listing 4.1 on Page 65. They are located in the functions `FlexFilterRowV()`, `FlexFilterColumnV()`, `GradientMagThreshold()`, `CalcWeights()`, `GenerateNewParticles()` and `GetObservation()`. All of these six loops are parallelized in PARSEC according to the doAll pattern. No loop other than these six is parallelized. This means, we reported neither false negatives nor false positives. ICC did not find any doAll loops in bodytrack.

Our approach can identify parallel patterns at different levels of the hierarchy in the program execution tree. Identification of a combination of more than one parallel pattern at the same level would need specific templates for each combination. That new templates need to be defined for each combination of patterns is a limitation of the current approach, which will be addressed in future work.

The correlation coefficient of doAll identification for the loop in Listing 4.1 was $0.98$. It was due to two loop-carried dependences, as shown in Figure 4.1(a) on Page 66. The first dependence was because $CU_{estimate}$ uses the values generated by $CU_{output}$ in the previous stage and the second was a self-dependence of $CU_{output}$. We could not resolve these dependences to make this loop a doAll loop. This shows that loop-carried dependences can sometimes not be resolved, even if their number is low. A correlation coefficient close to $1.0$ does not necessarily mean that the loop can be easily converted into a doAll loop.

**Blackscholes**

In blackscholes, we found two nested loops with a correlation coefficient of $1.0$, together covering
99.75% of the total read and write instructions. A closer investigation of the code revealed that
the outer loop is not an intrinsic part of blackscholes, but rather a wrapper in PARSEC which runs
the blackscholes algorithm a specified number of times on the same input. This is why it was
not parallelized in PARSEC, although it could have been in principle. The inner loop is shown in
Listing 4.7. It is the main blackscholes loop, which calculates the value of an option. The parallel
implementation in PARSEC follows the doAll pattern. ICC also found the same loops to be doAll loops.
We did not find any loop in blackscholes with a correlation coefficient of less than $1.0$. Listing 4.8
shows the parallelization of the main loop of blackscholes discussed earlier. This shows the simplicity
with which the doAll loops can be parallelized using OpenMP.

Listing 4.7: DoAll loop in blackscholes.

```
for (i=start; i<end; i++) {
    price = BlkSchlsEqEuroNoDiv( i, 0);
    prices[i] = price;
}
```

Listing 4.8: Parallelized loop of blackscholes.

```
#pragma omp parallel for private(i, price)
for (i=start; i<end; i++) {
    price = BlkSchlsEqEuroNoDiv( i, 0);
    prices[i] = price;
}
```

**Dedup**

We could not find any hotspot loops in dedup with a correlation coefficient of $1.0$. ICC also failed to
find any doAll loops in dedup. The two nested hotspot loops for which we implemented pipelines
received a correlation coefficient of $0.99$ and $0.94$, respectively.

The outer loop could not be converted into a doAll loop because of the serial input and one
variable causing loop-carried dependences. This dependence could not be resolved. Similarly, the
inner loop could not be converted to doAll due to the splitting of input buffer in each iteration
depending on the split from the previous iteration. Also the output of results in this loop was serial.

**LibVorbis**

The doAll search in LibVorbis produced results similar to the one for dedup. The two nested hotspot
loops that were identified as pipelines were also checked for doAll patterns and correlation coefficients
of $0.96$ and $0.70$ were computed. The outer loop had loop-carried dependences due to the serial file
input and some internal flags of the LibVorbis library. These dependences were not resolvable, as the
LibVorbis library uses these flags for maintaining its internal state. The inner loop also depended on
the internal flags of the LibVorbis library. Moreover, the file output was sequential. These dependences

could not be resolved to realize a doAll pattern in the loop.

**Others**

DoAll patterns were identified in the benchmarks *c-ray, md5, ray-rot, rotate, alignment,* and *nqueens.*
All of the above benchmarks were shipped with their parallel version in their respected benchmark
suites. We cross checked our identified doAll loops with the already available versions and found all
the doAll loops parallelized using the same pattern. We did not have any false negatives in any of
the cases evaluated.

## 4.4 Summary

In this chapter, we described how to infer potential parallelization patterns from sequential programs.
To this end, we adapted the idea of template matching, which is already employed to identify
sequential design patterns in UML diagrams. A specific challenge we addressed is – in comparison to
UML diagrams – the less rigid structure of parallel design patterns. Our approach identified those
parts of the code that can be parallelized and showed how to map these parts of code onto the pattern
structure, enabling a quick implementation of the pattern. We believe that this will provide significant
help for organizations facing the challenge of parallelizing large numbers of legacy programs.

The evaluation demonstrates the successful identification of pipeline and doAll parallel patterns
in different applications. This implies that template matching can be used successfully for the
identification of parallel patterns that exhibit specific structural characteristics like chain dependences
in pipelines and the absence of inter-iteration dependences in doAll loops.

More parallel patterns for which we can define template vectors can be easily added to our
framework in the future. In this way, the programmer may become able to choose among several
pattern alternatives for the same piece of code, taking their correlation coefficients into account.
Ultimately, however, our vision is the pattern-guided semi-automatic transformation of sequential
code into a parallel version.

# 5 Pattern Identification with Regression Analysis

The usage of template matching for the identification of parallel patterns has shown promising results, as discussed in Chapter 4. It can be used to identify parallel patterns with definite characteristics, such as chain dependences in a pipeline. The value of the correlation coefficient calculated during template matching furnishes the degree of match between the vectors of pattern and application. However, we realized that in some cases, applying template matching is not sufficient.

These are the cases where the value of correlation coefficient helps in identifying a pattern, but it does not provide any meaningful details about the efficiency of that pattern. The multi-loop pipeline pattern is one such example. In this pattern, a pipeline exists among the iterations of different loops. Template matching only identifies the pipeline among these loops, but it cannot determine whether the implementation of such a pipeline will be beneficial. Because it depends on which iteration of the dependent loop relies upon which iteration of the dependee loop. The details of the multi-loop pipeline pattern are discussed in Section 5.1.

To address this shortcoming, we extracted some more data from the input program. This data is used to compute the estimated efficiency of the pattern by employing a well-known statistical tool called simple linear regression analysis [114, 115]. Simple linear regression is used to predict the relationship among an independent variable and a dependent variable. The aim of this analysis is to discover a linear function that predicts the value of the dependent variable given the value of the independent variable with least error. It tries to find the line that best-fits the observed pairs $(X_i, Y_i)$. Linear regression helps in finding coefficients $a$ and $b$ of the linear equation:

$$Y = aX + b \tag{5.1}$$

The values of coefficients $a$ and $b$ help us in identifying a parallel pattern and its efficiency. We use simple linear regression analysis for the identification of multi-loop pipelines and reduction.

## 5.1 Multi-loop Pipeline

Multi-loop pipelines are special cases of pipelines which are hard to identify because they stretch across more than one loop. In this scenario, one iteration of a loop depends on one or more iterations of a preceding loop in a sequential application. This can be easily transformed into a pipeline by mapping iterations of different loops onto different stages of the pipeline. In Listing 5.1, we show an example code where a multi-loop pipeline exists. Every $i_{th}$ iteration of the later loop depends on the $i_{th}$ iteration of the former loop.

Such a pipeline can be identified using template matching by looking at dependences among different loops, instead of CUs of a single loop. However, in this case, we cannot determine the efficiency of the pipeline identified using the value of the correlation coefficient. For example, it may happen that the very first iteration of the dependent loop depends on the last iteration of the previous loop. Here, the value of correlation coefficient will be 1.00, but in practice such a pipeline, if implemented, will be highly inefficient. For this reason, we use simple linear regression analysis to determine the efficiency of these pipelines.

Listing 5.1: Example of multi-loop pipeline.

```
for (. . .)// Loop x
    a[i] = foo(i);

for (. . .)// Loop y
    b[i] = bar(a[i]);
```

## Identification Approach

We developed a mechanism that automatically identifies multi-loop pipelines. Algorithm 3 shows the algorithm of multi-loop pipeline identification. All pairs of hotspot loops (in which one loop is data dependent on the other) are gathered from the PET of a serial program. An LLVM-pass instruments only the load and store instructions that create these dependences. The pass records the iteration numbers of the loops for these instructions.

$PEG = getExecutionGraph(serialProgram)$
$loops = getHotspotLoops(PEG)$
$loopPairs = findLoopsWithDependence(loops)$
**for each** $pair$ **in** $loopPairs$ **do**
    $instructions = findDepCreatingInstructs(pair)$
    $instrumentInstructions(instructions, serialProgram)$
**end**
$iterInfo = runProgram(serialProgram)$
**for each** $pair$ **in** $loopPairs$ **do**
    $cIterPairs = identifyConflictingIters(pair, iterInfo)$
    $Result[pair] = computeRegression(cIterPairs)$
**end**
$return\ Result$

**Algorithm 3** : Identification of multi-loop pipeline.

The instrumented program is then executed with representative inputs; all the recorded information about the iteration numbers and memory addresses are dumped into an output file. Suppose, there is a dependence between loop $x$ and loop $y$ (as in Listing 5.1) where loop $x$ writes to and loop $y$ reads from the same memory locations. A post-analysis filters out the last iteration number $i_x$ of loop $x$ that wrote to a memory location $m$ and the first iteration number $i_y$ of loop $y$ that read from the same memory location $m$. These filtered iteration numbers are stored in pairs $(i_x, i_y)$ for each

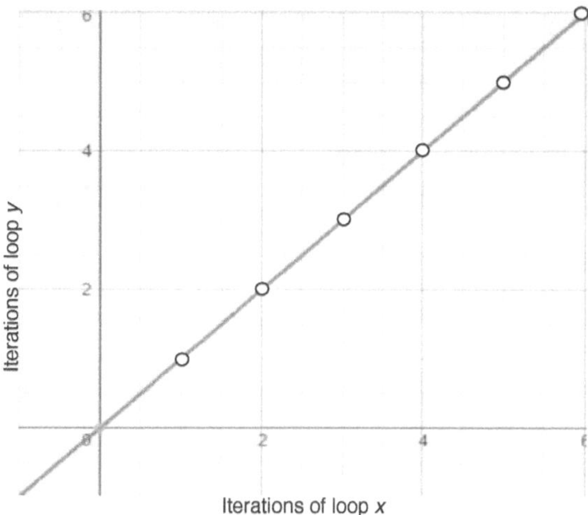



Figure 5.1: Regression line for a perfect multi-loop pipeline.

loop pair identified in the previous step. Each pair denotes that iteration $i_y$ of loop $y$ is dependent on iteration $i_x$ of loop $x$.

To estimate the relationship between pairs of iterations, we use simple linear regression analysis. As mentioned earlier, it estimates the relationship between a dependent variable $Y$ and an independent variable $X$. We estimate coefficients $a$ and $b$ of the linear equation shown in Equation 5.1. Once we have these coefficients, we can estimate the potential existence of a multi-loop pipeline between two loops. It is a perfect multi-loop pipeline when each $i_{th}$ iteration of loop $y$ depends exactly on the $i_{th}$ iteration of loop $x$. The regression line of a perfect multi-loop pipeline is shown in figure 5.1. The values of coefficients $a$ and $b$ for this line are 1 and 0, respectively.

We denote the area under the line of a perfect pipeline as $\int_{perfect}$. Similarly, linear regression gives us a regression line for the relationship between the iterations of loops $x$ and $y$. The area below the generated regression line is denoted by $\int_{current}$. To estimate the efficiency of multi-loop pipeline between the iterations of loops $x$ and $y$, we calculate the multi-loop pipeline efficiency factor $e$ with formula:

$$e = \frac{\int_{current}}{\int_{perfect}} \tag{5.2}$$

The value of $e$ represents how efficient the multi-loop pipeline will be. If the value of $e$ is equal to 1.0, then we have a perfect multi-loop pipeline. If the value of $e$ is close to or equal to 0, the multi-loop pipeline will be inefficient; the second loop $y$ will have to wait for almost all iterations of the first loop $x$ to finish before any iteration of loop $y$ can start. If $e$ is much higher than 1, there is a possibility of running both loops almost in parallel to each other with minimal synchronization needed between their iterations. In addition to the efficiency factor $e$, the coefficients $a$ and $b$ of Equation 5.1 give some insights into the implementation of a multi-loop pipeline. The impact of the

Table 5.1: Impact of the values of coefficients $a$ and $b$ in Equation 5.1 on multi-loop pipelines (©2016 IEEE).

| Coefficient | Value | Description |
|---|---|---|
| $a$ | 1 | 1 iteration of loop $y$ depends exactly on one iteration of loop $x$. |
| $a$ | $< 1$ | 1 iteration of loop $y$ depends on $1/a$ iterations of loop $x$. |
| $a$ | $> 1$ | $a$ iterations of loop $y$ depend on 1 iteration of loop $x$. So $a$ iterations of loop $y$ can be executed after 1 iteration of loop $x$. |
| $b$ | 0 | All iterations of loop $y$ depend on all iterations of loop $x$. |
| $b$ | $< 0$ | No iteration of loop $y$ depend on first $b$ iterations of loop $x$. |
| $b$ | $> 0$ | The first $b$ iterations of loop $y$ do not depend on any iteration of loop $x$. |

values of $a$ and $b$ on a multi-loop pipeline are shown in Table 5.1. With the help of all three values ($e$, $a$, and $b$), programmers can easily predict the practicality of multi-loop pipelines. Listing 5.2 shows the format of the output generated by the multi-loop pipeline identification process. The loop identifiers are assigned by DiscoPoP during the profiling. The efficiency factor and the coefficients for each identified multi-loop pipeline are reported for each loop pair.

Listing 5.2: Output format of multi-loop pipeline identification.

```
1    LoopID1    LoopID2:    e    (a, b)
```

If there is a chain of more than two loops depending on each other (e.g., loop $y$ depends on loop $x$ and loop $z$ depends on loop $y$), our tool separately provides the relationships between loops $x$ and $y$ as well as loops $y$ and $z$. If there is a chain dependence of $n$ loops, it gives $n$ pairs of relationships. A pipeline of $n$ stages can be easily implemented by merging the information provided by the tool. Moreover the loops in each stage of a multi-loop pipeline may be parallelized using other parallel patterns such as doAll, reduction or pipeline etc.

## 5.2 Fusion

In some cases, an identified multi-loop pipeline can be optimized using fusion. Fusion is a loop optimization technique used by compilers to merge two loops into a single one if they iterate over the same range [24]. This is done to reduce loop overhead and increase granularity.

Fusion done by compilers is limited by static analysis and only effective if the loops are next to each other. Our approach suggests the fusion of loops based on dynamic analysis and the suggested loops may be lexically apart from each other in the actual source code.

### Identification Approach

We use the data gathered during our analysis of multi-loop pipelines to identify a fusion. A fusion of loops $x$ and $y$ can occur if:

- both loops $x$ and $y$ are doAll loops.

- the values of coefficients $a$ and $b$ of Equation 5.1 are exactly 1 and 0, respectively. As a result, the efficiency factor $e$ from the multi-loop analysis will be 1.

Both conditions ensure that the fused loop will not contain any loop-carried dependences and can be parallelized using doAll. If these conditions are met, we suggest fusing them together as a single loop and parallelizing it with doAll. This reduces the synchronization overhead for each loop and makes the parallelization more coarse-grained.

Another advantage of employing fusion is that it improves locality of reference [24]. DiscoPoP currently does not report the amount of data being handled by a single loop iteration. Therefore, our approach does not consider locality of reference when suggesting loop fusion. We will address this feature in our future work.

## 5.3 Reduction

The reduction pattern enables programmers to parallelize loops with a specific type of inter-iteration dependence. It can only be used when a loop uses an associative binary operator to reduce all elements of a container to a single scalar value, e.g., a loop summing all the elements of an array. State-of-the-art compilers generally recognize reduction, though pointer aliasing and array referencing may make them miss some reduction opportunities [103]. Our dynamic approach overcomes these limitations.

---

**Input :** loopID
$Vars = getAllWrittenVars()$
**for** *each* $Var$ **in** $Vars$ **do**
    $writeLines = getWriteLines(Var)$
    **if** $|writeLines| \; ! = \; 1$ **then**
      |  *next*
    **end**
    $readLines = getReadLines(Var)$
    **if** $|readLines| \; ! = 1 \; || \; readLines \; ! = \; writeLines$ **then**
      |  *next*
    **end**
    $Result[loopID] + = $ Reduction candidate $Var$ at $writeLines$.
**end**
*return Result*

---

**Algorithm 4 :** Reduction identification (©2016 IEEE).

### Identification Approach

We identify reduction using the same LLVM-pass we developed for multi-loop pipelines. The pass instruments all LLVM-IR instructions creating inter-iteration dependences in a loop. However, there are two differences between reduction and multi-loop pipeline analysis: 1) dependence analysis between the iterations is done for iterations of the same loop instead of two separate loops; and 2) source line numbers are also recorded for each write and read operation for each variable involved.

Currently, we can identify only the simplest case of reduction. In Algorithm 4, we show the steps for the identifying a reduction pattern. All variables accessed by instructions, instrumented during the instrumentation phase, are checked for reduction. If a memory address is written only on a single source line of a loop and read only at the same source line, the loop is reported as a possible candidate for a reduction pattern. The identification results also contain source line numbers where reduction may occur.

Our approach does not automatically identify the operator used at the source line number reported for a possible reduction. Currently, this burden is still left to the programmer, who must decide whether the operation in the reported source line number can be parallelized using reduction.

## 5.4  Geometric Decomposition

The geometric decomposition pattern is used whenever parallelism is driven by dividing data and working on it simultaneously using different threads. In a serial program, we often need to call the same program code over a huge amount of data, motivating the term *single program multiple data* (SPMD). Such a paradigm can be easily parallelized using geometric decomposition if the data can be divided into chunks and processed independently by separate threads.

### Identification Approach

DoAll is a case of geometric decomposition in which the iterations of a loop run independently of each other on separate data. However, the doAll pattern is restricted to loops. Programmers may overlook geometric decomposition opportunities on the function level. In our approach to the identification of the geometric decomposition, we gather all immediate child nodes of a hotspot function node from the PEG. We analyze all the loops in the current function and all the loops in the functions called directly from the current function. If all the analyzed loop nodes are either doAll or reduction loops, then we suggest this function as a candidate for geometric decomposition. The identification process of the geometric decomposition pattern is shown in Algorithm 5.

DiscoPoP cannot determine the amounts and types of data being supplied to a function. Currently, the programmer must decide how the data entered into the function can be split into separate chunks so that the same function can be called separately for each chunk of data in separate threads. This often facilitates speedup; geometric decomposition coarsens the granularity of parallelization and reduces synchronization overhead as compared to the implementation of a separate doAll or reduction pattern for each individual loop in the function.

```
Function identifyGD(PET, FuncNode)
    Nodes = getImmediateChildren(PEG, FuncNode)
    for each child in Nodes do
        if child is Loop then
            if !doAllLoop(child) OR !reductionLoop(child) then
                return false
            end
        end
        else if child is Function then
            if !AllLoopsDoallOrReduction(child) then
                return false
            end
        end
    end
    return true
```

**Algorithm 5** : Identification of geometric decomposition (©2016 IEEE).

## 5.5 Evaluation

We evaluated our approach with four different benchmark suites. Starbench consists of C/C++ benchmarks from diverse fields such as image processing, hashing, compression and so on [111]. The Barcelona OpenMP Task Suite (BOTS) [112] has been designed to study task parallelism in OpenMP. The Polyhedral Benchmark suite (Polybench)) [30] consists of benchmarks with kernels from areas such as data mining and linear algebra etc. We also used another benchmark named *fluidanimate* from the Parsec benchmark suite [109] to test multi-loop pipeline identification. Starbench, BOTS, and Parsec include sequential as well as parallel versions of these benchmarks, allowing us to compare our identification results with the available parallel versions in these suites. We implemented all those identified patterns ourselves that were not found in the parallel versions of the respective benchmarks.

Polybench does not have any parallel versions included in the suite itself. Grauer-Gray et al. [116] reported the parallelization of some benchmarks from Polybench. Their target platform for the parallel versions are GPUs. We compared pattern identification results with their reported list of benchmarks and found that they only listed those benchmarks where we identified the doAll, reduction or fusion patterns. This comparison validates our approach because all of these patterns are suitable for GPU-like architectures. We implemented all the identified patterns in polybench by manually transforming the sequential versions. We explicitly mention whenever we implemented an identified pattern on our own. The comparison of our achieved speedups for the parallelized versions of Polybench benchmarks with the ones reported by Grauer-Gray et al. is not possible because our implementation was targeted for general purpose CPUs.

We conducted our evaluation on a machine with 2×8-core Intel Xeon E5-2650 2 GHz processors with hyper-threading and 32 GB memory, running Ubuntu 12.04 (64-bit server edition). We compiled

Table 5.2: Overall pattern identification results for different applications from Starbench, BOTS, Polybench and Parsec (©2016 IEEE).

| Application | Benchmark Suite | LOC | Exec Inst % in Hotspot | Speedup | Threads | Identified Pattern |
|---|---|---|---|---|---|---|
| ludcmp | Polybench | 135 | 88.64% | 13.58 | 32 | Multi-loop pipeline |
| reg_detect | Polybench | 137 | 99.50% | 2.26 | 16 | Multi-loop pipeline |
| fluidanimate | Parsec | 3987 | 99.54% | 1.5 | 3 | Multi-loop pipeline |
| rot-cc | Starbench | 578 | 94.53% | 16.18 | 32 | Fusion |
| Correlation | Polybench | 137 | 99.27% | 7.09 | 32 | Fusion |
| 2mm | Polybench | 153 | 99.19% | 5.41 | 32 | Fusion |
| nqueens | BOTS | 118 | 100.00% | 8.38 | 32 | Reduction |
| bicg | Polybench | 191 | 74.58% | 5.64 | 8 | Reduction |
| gesummv | Polybench | 188 | 65.33% | 5.06 | 8 | Reduction |
| kmeans | Starbench | 347 | 2.04% | 3.97 | 8 | Geometric decomposition + Reduction |
| streamcluster | Starbench | 551 | 49.99% | 6.38 | 32 | Geometric decomposition |

Table 5.3: Summary of multi-loop pipeline identification (©2016 IEEE).

| Application | a | b | e |
|---|---|---|---|
| ludcmp | 1 | 0 | 1 |
| reg_detect | 1 | −1 | 0.99 |
| fluidanimate | 0.05 | −3.50 | 0.97 |

all the benchmarks with Clang 3.3 using `-g -O2` flags. We report the overall identification results of all these benchmarks in Table 5.2. The table shows lines of code, percentage of executed instructions in the hotspot, speedup, maximum number of threads to achieve that speedup, and the name of the identified parallel pattern. We tested all these benchmarks with a maximum of 32 threads and we report the number of threads where the benchmarks achieved their highest speedup. Multiple runs of a benchmark were done and the results were merged, whenever multiple imputs for that benchmark were available, to mitigate the limitations of the dynamic analysis. In the following subsections, we will discuss the results of each identification approach in detail.

### 5.5.1 Multi-loop Pipeline

Our tool identified multi-loop pipelines in three benchmarks. *ludcmp*, *reg_detect*, and *fluidanimate* had multi-loop pipelines. Table 5.3 shows the values for the coefficients $a$ and $b$ as well as the efficiency factor $e$ for the identified multi-loop pipelines.

In *ludcmp*, we found a multi-loop pipeline between the two loops in function `kernel_ludcmp()`.

Listing 5.3 shows the simplified code of these loops. The first loop was a doAll loop; the second loop had inter-iteration dependences. We found a one-to-one dependence between iterations of the two loops so it was reported as a perfect multi-loop pipeline.

Listing 5.3: Hotspot loops of *ludcmp* with a multi-loop pipeline identified between them.

```
for (i = 0; i < _PB_N; i++){
    //Write to array A
}
y[0] = b[0];
for (i = 0; i < _PB_N; i++){
    w = b[i];
    //read from array A and write to w
    y[i] = w;
}
```

Listing 5.4: Multi-loop pipeline implementation of the *ludcmp* loops shown in Listing 5.3.

```
y[0] = b[0];
#pragma omp parallel num_threads(NT) private(w)
{
    #pragma omp single
    for (int block=0; block < numOfBlocks; block++)
    {
        #pragma omp task depend(out: inter_flags[block])
        {
            for (int i=block*BLOCK_SIZE+1; i<min(_PB_N,
(block+1)*BLOCK_SIZE+1); i++) {
                //Write to array A
            }
        }
    }
    #pragma omp task depend(in: inter_flags[block], intra_flags[block])
depend(out: intra_flags[block+1])
    {
        for (int i=block*BLOCK_SIZE+1; i<min(_PB_N,
(block+1)*BLOCK_SIZE+1); i++) {
            w = b[i];
            //read from array A and write to w
            y[i] = w;
        }
    }
}
```

We implemented the identified multi-loop pipeline and achieved a maximum speedup of 13.58 with 32 threads. Listing 5.4 shows the parallel implementation of the identified multi-loop pipeline. The implementation was done using the *depend* and *task* clauses of OpenMP. Figure 5.2 shows a

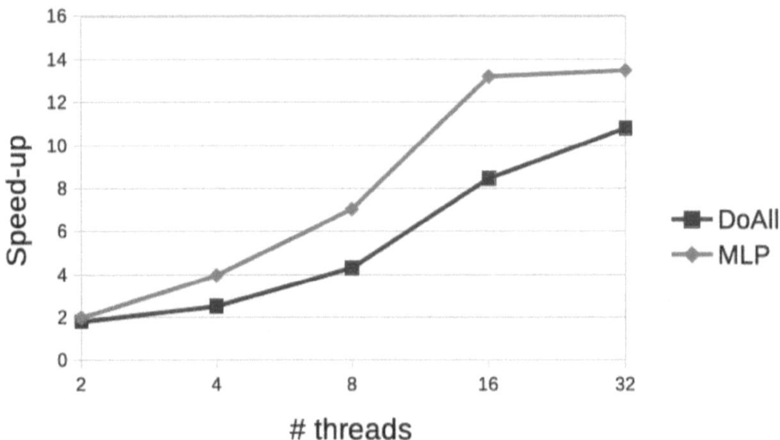


Figure 5.2: Speedups achieved for doAll and multi-loop pipeline implementations of *ludcmp*.

comparison between the doAll and the multi-loop pipeline implementations of ludcmp. In the doAll case, only the first loop was parallelized using the OpenMP *for* construct. Our multi-loop pipeline version of *ludcmp* always achieved better speedup than the doAll version. This shows the advantage of mixing more than one parallel pattern in terms of performance.

Listing 5.5: Hotspot loops of *reg_detect* with a multi-loop pipeline identified between them.

```
1   void kernel_reg_detect(){
2       ...
3       for (i=0; i<PB_MAXGRID-1; i++) {
4           ...
5           mean[i][j] = ....
6       }
7       for (i=1; i<PB_MAXGRID-1; i++) {
8           ...
9           path[i][j] = path[i-1][j-1] + mean[i][j]
10      }
11  }
```

In *reg_detect*, we found a multi-loop pipeline between the two loops in function `kernel_reg_detect()`. Again, the first loop was a doAll loop and the second loop had inter-iteration dependences. Interestingly, the value of the coefficient $b$ was $-1$; no iteration of the second loop had any dependence on the first iteration of the first loop. We manually analyzed the code and confirmed this was indeed the case. Listing 5.5 shows the skeleton of the two loops from *reg_detect*. The value of $e$ was slightly affected by the value of coefficient $b$. We implemented a multi-loop pipeline for *reg_detect* by peeling the first iteration of the first loop. The remaining iterations of both

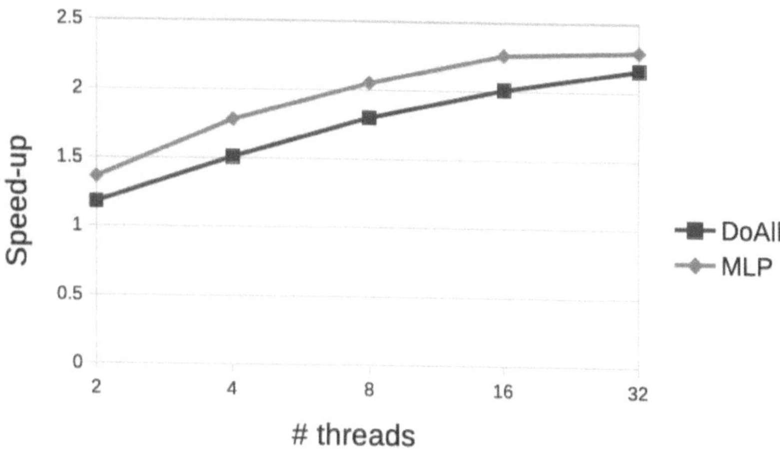

Figure 5.3: Speedups achieved for doAll and multi-loop pipeline implementations of *reg_detect*.

loops had a one-to-one dependence.

We implemented the multi-loop pipeline for these loops in the same way as for the *ludcmp* case. Our parallel version achieved a maximum speedup of 2.26 with 16 threads. Figure 5.3 shows a comparison between the speedups achieved for the doAll and multi-loop pipeline implementations of *reg_detect*. The results are similar to those of ludcmp in Figure 5.2, with the multi-loop pipeline version achieving better speedups than the doAll version.

In `ComputeForces()` of *fluidanimate*, we found two hotspot loops with a multi-loop pipeline between them. Listing 5.6 shows the pseudo code of these two loops at lines 3 and 5. Neither of these two loops were doAll loops. The values for $e, a,$ and $b$ were recorded as 0.97, 0.05, and $-3.50$, respectively. Based on our description of coefficients in Table 5.1, one iteration of the second loop depends on 20 iterations of the first loop (i.e. $1/a = 1/0.05 = 20$). We manually analyzed the loops and found that each iteration of the first loop updated the densities of the current cell and neighboring cells. The second loop reads and (again) updates the densities of these cells, confirming the results of our technique.

Listing 5.6: Hotspot loops of *fluidanimate* with a multi-loop pipeline identified among them.

```
1   for(i=0;i<num; ++i)
2       neigs[i] = get_neigs(cells[i]);
3   for{i=0; i<num; ++i}
4       ComputeDensities(cells[i], neigs[i]);
5   for{i=0; i<num; ++i}
6       ComputeForces(cells[i], neigs[i]);
```

We implemented a multi-loop pipeline for fluidanimate according to our identification results.

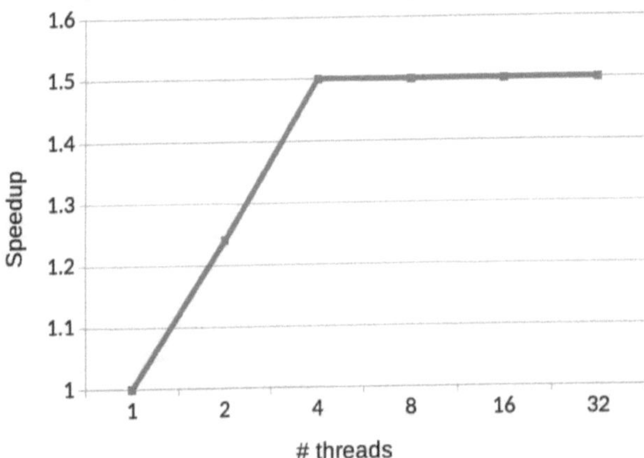

Figure 5.4: Speedups achieved for identified multi-loop pipeline in fuidanimate.

Due to the complex dependences between the loops, we achieved a maximum speedup of only $1.5$ with 3 threads. Figure 5.4 shows the speedups achieved for the implemented multi-loop pipeline for numbers of threads. Note that the speedup ceases to increase beyond three threads. This is due to the inherent structures of pipelines. The parallel version of *fluidanimate* in Parsec was implemented using a grid of parallel tasks with locks (stencil) to handle the complex dependences between the loops shown in Listing 5.6.

Listing 5.7: Hotspot loops of *correlation* benchmark suitable for fusion.

```
for(j = 0; j < m; j++){
    mean[j] = 0.0;
    for{i = 0; i < n; i++}
        mean[j] += data[i][j];
    mean[j] /= float_n;
}

for(j = 0; j < m; j++){
    stddev[j] = 0.0;
    for(i = 0; i < n; i++)
        stddev[j] += (data[i][j] - mean[j]) * (data[i][j] - mean[j]);
    stddev[j] /= float_n;
    stddev[j] = sqrt_of_array_cell(stddev, j);
}
```

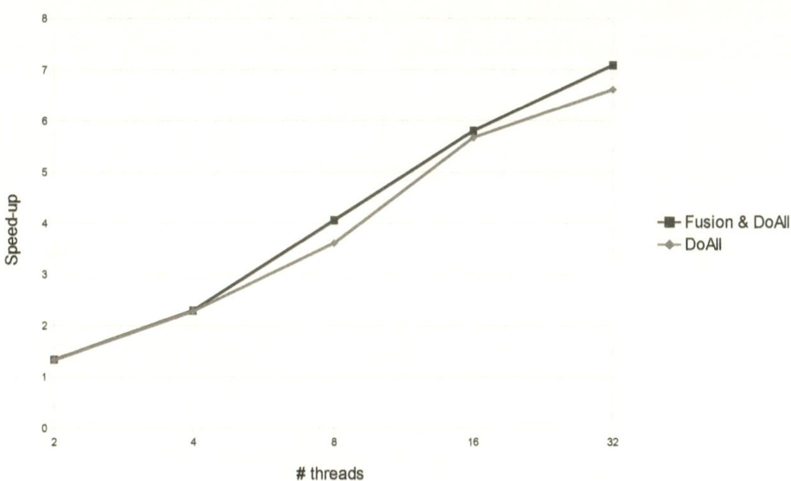

Figure 5.5: Comparison of speedups achieved for the fusion + doAll and doAll implementations of *correlation*.

## 5.5.2 Fusion

As mentioned in Section 5.2, fusion is an optimization performed over identified multi-loop pipelines. Hence all the cases reported here are also valid multi-loop pipelines. We found three cases of fusion in the benchmarks *rot-cc, Correlation,* and *2mm*.

To illustrate the kind of loops suitable for fusion, we show the loops identified for fusion from the *correlation* benchmark in Listing 5.7. The first loop at Line 1 is calculating means of every column of the input `data` matrix. The second loop at Line 8 computes standard deviations for each column. It consumes the means calculated in the first loop, hence creating a dependence. Both of these loops have the same iteration space and their iterations have a one-to-one dependence among them. Therefore, both these loops can be optimized using fusion.

*correlation* and *2mm* are from Polybench, they are not shipped with parallel versions. We implemented two parallel versions of these two benchmarks. One version was implemented with fusion and the other with doAll. Comparison of the speedups achieved for correlation and 2mm are shown in Figures 5.5 and 5.6 , respectively. Our results show that the versions with fusion achieve slightly better speedups compared to the doAll versions in both cases. One possible reason for this can be the use of memory cache, as the data loaded by the iteration of the first loop is directly used by the iteration of the second loop, hence avoiding the offloading of data from cache. *Rot-cc* from Starbench has a parallel version available, which was implemented by fusing the same two hotspot loops that our tool had identified.

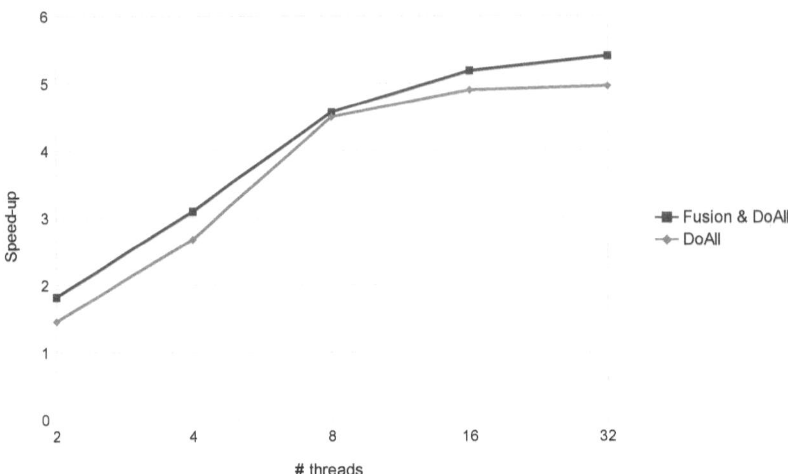

Figure 5.6: Comparison of speedups achieved for the fusion + doAll and doAll implementations of *2mm*.

### 5.5.3 Reduction

We identified reduction patterns in four benchmarks: *nqueens, kmeans, bicg,* and *gesummv*. In *nqueens*, we identified a reduction pattern in the main loop in function `nqueens()`. The existing parallel version of this benchmark in BOTS is implemented with reduction. The reduction loop identified in *kmeans* was inside a function suitable for geometric decomposition that will be discussed later.

Listing 5.8: Reduction candidate loop of *gesummv*.

```
for (i = 0; i < n; i++){
    tmp[i] = 0;
    y[i] = 0;
    for (j = 0; j < n; j++){
        tmp[i] = A[i][j] * x[j] + tmp[i];
        y[i] = B[i][j] * x[j] + y[i];
    }
    y[i] = alpha * tmp[i] + beta * y[i];
}
```

Our tool identified two variables candidate for reduction in *gesummv*. The inner loop at Line 4 shown in Listing 5.8 has variables `tmp[i]` and `y[i]` as reduction candidates. Both *bicg* and *gesummv* are from Polybench; therefore, we implemented their parallel versions manually via reduction. We achieved peak speedups of 5.64 for *bicg* and 5.06 for *gesummv* using 8 threads.

Table 5.4: Comparison of reduction identification results (©2016 IEEE).

| Tool | *nqueens* | *kmeans* | *bicg* | *gesummv* | *sum_local* | *sum_module* |
|---|---|---|---|---|---|---|
| Sambamba | NA | NA | ✓ | ✓ | ✓ | ✗ |
| icc | ✗ | ✗ | ✗ | ✗ | ✓ | ✗ |
| DiscoPoP | ✓ | ✓ | ✓ | ✓ | ✓ | ✓ |

As we pointed out earlier, state of the art compilers use static analysis to identify reduction pattern in loops. To demonstrate the advantage of our approach, we implemented two synthetic benchmarks we call *sum_local* and *sum_module*. We show the code of these synthetic benchmarks in Listings 5.9 and 5.10, respectively. The reduction operation is done in the lexical extent of the loop in *sum_local*; in *sum_module*, the reduction operation is done in another function called from within the loop.

Listing 5.9: Code of *sum_local* reduction.

```
1   int sum_local(int* arr, int size){
2       int sum=0;
3       for (i = 0; i < size; i++)
4           sum += arr[i];
5   }
```

Listing 5.10: Code of *sum_module* reduction.

```
1    int sum_module(int &sum, int val){
2        int x = ...//do some heavy work on val
3        d += x;
4        return x;
5    }
6    int sum_module(int* arr, int size){
7        int sum=0;
8        for (i = 0; i < size; i++){
9            x = sum_module(sum, arr[i]);
10           foo(x);
11       }
12       return sum;
13   }
```

We compared our results for these two synthetic benchmarks with Intel's *icc* compiler [103] and *Sambamba* [86], another tool that statically identifies reductions. The results are shown in Table 5.4. *icc* was able to identify the reduction only in *sum_local*. On the other hand, *Sambamba* was not able to identify the reduction pattern in *sum_module*. Both *icc* and *Sambamba* could not identify reduction in *sum_module*. The limitations of static analysis prevented both *icc* and *Sambamba* from identifying the reduction in *sum_module*. In contrast, our approach identified the reduction patterns in both benchmarks.

### 5.5.4 Geometric Decomposition

We identified geometric decomposition in *streamcluster* and *kmeans* from Starbench. The main loop in function `streamCluster()` of *streamcluster* is shown in Listing 5.11. It creates new clusters in each iteration and cannot be parallelized with doAll or pipeline because it uses the new clusters formed at the end of the last iteration as an input to the next iteration. Therefore, we identified no parallel pattern in `streamCluster()`. The next hotspot was function `localSearch()` being called within the biggest hotspot loop.

All the loops in `localSearch()` and the loops in the functions called by `localSearch()` were identified as doAll or reduction loops. We did not report the loops we identified as reduction in Table 5.2 because they are not hotspots. Our tool recommended `localSearch()` as a possible candidate for geometric decomposition. We analyzed the parallel version of *streamcluster* from Starbench and found that it was parallelized via the geometric decomposition of function `localSearch()`. The pseudo code of the parallel *streamcluster* implementation is shown in Listing 5.12. Similarly, we reported the function `cluster()` of *kmeans* as a possible geometric decomposition candidate; this finding was confirmed by the existing parallel version of *kmeans*.

Listing 5.11: Hotspot loop of *streamcluster* benchmark.

```
1    void streamCluster(){
2        while(1){
3            ...
4            localSearch(points);
5            ...
6            if (eof)
7                break;
8    }}
```

Listing 5.12: Parallel implementation of *streamcluster* benchmark.

```
1    void streamCluster(){
2        while(1){
3            ...
4            for(i=0; i<num_chuncks; i++)
5                new_thread(localSearch( points[i*chunk_size], chunk_size));
6            ...
7            if (eof)
8                break;
9    }}
```

## 5.6 Summary

The successful use of template matching to identify parallel patterns was established in Chapter 4. But for some patterns, like multi-loop pipelines, the results of identification through template matching are not enough to estimate their efficiency. For this purpose, we need extra information from the analyzed program.

We developed an approach to extract the missing information through an LLVM pass. This pass uses the dependences output by DiscoPoP. It instruments only specific reads and writes that create dependences among the iterations of different loops. The data gathered from this instrumentation is analyzed with simple linear regression analysis. This analysis helps in identification of such patterns and also provides an estimation of their efficiency. In this chapter, we discussed in detail the identification approach of four patterns namely: multi-loop pipelines, fusion, reduction, and geometric decomposition.

The evaluation of our approach shows promising results. We identified multi-loop pipelines in three benchmarks with different efficiency factors. Our results show that the benchmark with the maximum expected efficiency (i.e., $1.00$) also gained most speedup ($13.58\times$) out of the three with $32$ threads on a $16$ core machine. Fusion is a loop optimization that can be applied over identified multi-loop pipelines. A multi-loop pipeline between two loops with the same iteration space and no inter-iteration dependences can be optimized using fusion. We were able to achieve a maximum speedup of $16.18\times$ for one of the three fusion candidate benchmarks on the same machine as mentioned above. We also showed that for two of the cases, fusion achieved slightly better speedup compared to separate doAll implementations of the involved loops.

Reduction was identified in the hotspots of four benchmarks. The comparison of our reduction identification with other state-of-the-art approaches shows that our approach is better. We not only identify reduction in the lexical context of a loop but also in its dynamic context. Geometric decomposition was identified in two benchmarks. We suggest a geometric decomposition candidate function, if all of its loops can be parallelized using doAll or reduction patterns. These candidate parallelization opportunities maybe overlooked by the programmer as the identification process is not straightforward.

Additionally, the evaluation shows that our pattern identification framework is able to identify multiple parallel patterns in a sequential program even at different levels. In the future, we want to improve our reduction identification so that we can automatically infer the type of reduction operator and cover more complex reduction scenarios. We also plan to support more parallel patterns and loop optimizations, such as peeling and fission, to enhance our parallel pattern identification framework.

# 6 Pattern Identification with Breadth-First Search

Template matching requires a pattern vector to match against, in order to identify that pattern. Sometimes, it becomes very difficult to construct a pattern graph or a pattern vector, if the parallel pattern to be matched is of a dynamic nature. For example, in the task parallelism pattern, the number of parallel tasks and dependences among them is exceedingly dependent on the structure of the application under study. So, the lack of some definite characteristics, like chain dependences in a pipeline, and the highly dynamic structure make the extraction of a template vector for the task parallelism pattern impossible.

Therefore, template matching cannot be applied to the identification of the task parallelism pattern. For this reason, we decided to employ a well-know algorithm called *breadth-first search* (BFS) [117]. It is a very simple technique for graph exploration. The algorithm starts searching from a source node and traverses all the reachable nodes in the graph in a specific order.

Algorithm 6 shows the pseudo code of standard BFS. The inputs are a graph $G$ and a starting node $S$. The starting node is labeled as discovered and pushed into a queue. A loop iterates over the nodes in the queue until it is empty. Inside this loop, a node is popped out in each iteration. If it is the desired node it is returned and the algorithm stops. Otherwise, all the adjacent nodes of the current node are traversed and checked for labeling. If they are not labeled yet, then they are labeled as discovered and pushed into the queue.

The BFS algorithm traverses the adjacent nodes of the source node first before going to the next depth level. This is why it is called breadth-first. It is one of the basic graph traversal algorithms. We use a slightly modified version of the BFS algorithm for the identification of the task parallelism pattern.

## 6.1 Task Parallelism

The task parallelism pattern has been discussed in Chapter 2. This pattern exploits parallelism in an application based on the concurrent execution of tasks. A task is usually a program construct like a loop or a function.

As discussed in Chapter 2, the iterations of a loop can also be considered as separate tasks and can be parallelized. A special variant of the task parallelism pattern is the doAll pattern. In this pattern, all the iterations of a loop are independent. We have already discussed in detail the identification mechanism of the doAll pattern in Chapter 4. The lack of any dependence constraints in the doAll pattern makes it simple to construct a template vector for its identification.

On the other hand, dependences play a major role in general task parallelism. They effect the scheduling of parallel tasks, as all the dependences of a task have to be fulfilled prior to its execution

```
Function BFS(graph G, node S)
    S.label = discovered
    Queue.push(S)
    while !Queue.empty() do
        Node n = Queue.pop()
        if n is required node then
            return n
        end
        foreach Node p adjacent to n do
            if p.label != discovered then
                p.label = discovered
                Queue.push(p)
            end
        end
    end
    return NULL
```

**Algorithm 6 :** Pseudo code of breadth-first search.

or there must be some synchronization mechanism employed among dependent parallel tasks. The dependences among the tasks are subject to the structure of the program and cannot be generalized. Due to this reason, there cannot be any representative template vector for the task parallelism pattern.

### Identification approach

The task parallelism pattern consists of a collection of concurrent independent tasks. The most straightforward method of finding task parallelism is to look for totally independent CUs in the PEG of a region and consider them as independent tasks. However, most cases include dependences among CUs. Previous task parallelism identification techniques [38, 57] were limited; they did not report the appropriate synchronization among the identified parallel tasks. We use *breadth-first search* (BFS) to identify task parallelism in the PEG of a region and classify the identified tasks in a way that helps programmers to easily synchronize them in parallel execution.

Algorithm 7 illustrates our task parallelism identification approach. We start with the first unmarked CU in the PEG of a hotspot in serial order of execution and mark it as a *fork* CU. All unmarked CUs dependent on the current CU are marked as *worker* CUs. A dependent CU is updated to a *barrier* CU if it was already marked. A barrier CU depends on more than one CU. A worker or barrier CU may behave as a fork CU for all of its dependent CUs. Once all CUs traversable by a current fork CU have been successfully marked, the algorithm searches for any unmarked CUs in the PEG. If one is found, this CU is marked as a fork point and the entire process described above is repeated for its dependent nodes.

The output of our task parallelism identification algorithm is a classification of CUs into fork,

```
Function identifyTP(PEG of a hotspot region)
    while !AllCUsMarked(PEG) do
        S = getFirstUnmarkedCU(PEG)
        S.mark = "Fork"
        Queue.push(S)
        while !Queue.empty() do
            N = Queue.pop()
            foreach D in N.dependents do
                if D.mark is NULL then
                    D.mark = "Worker"
                end
                else
                    D.mark = "Barrier"
                end
                Queue.push(D)
            end
        end
    end
    checkParallelBarriers(PEG)
```

**Algorithm 7** : Identification of task parallelism (©2016 IEEE).

worker, and barrier CUs. It contains precise details about which CUs fork which worker CUs and which CUs are barrier CUs for these workers. In Figure 6.1, we show the PEG of function `cilksort()` of the *sort* application from the BOTS benchmark suite [112] and the classification of CUs by our algorithm. The output of our algorithm for this PEG shows that $CU_0$ forks four workers: $CU_1, CU_2, CU_3$ and $CU_4$. $CU_5$ is a barrier for workers $CU_1$ and $CU_2$. $CU_6$ is a barrier for workers $CU_3$ and $CU_4$. Finally, $CU_7$ is a barrier for $CU_5$ and $CU_6$. To see if two barriers can run parallel to each other, we check for a directed path from one barrier to the other in the PEG (or vice versa). The absence of a path denotes that two barriers can run in parallel; otherwise, they cannot. For example, there is no directed path between $CU_5$ and $CU_6$. Hence, they can run in parallel. On the other hand, there are paths from $CU_7$ to $CU_5$ and $CU_6$. So, $CU_7$ cannot run in parallel with either of these two barrier CUs.

We calculate the *estimated speedup* metric for the task parallelism we identified. This is calculated by dividing the total number of LLVM-IR instructions in the hotspot by the total number of LLVM-IR instructions on the critical path of that hotspot. This metric, to some extent, gives an indication of the effectiveness of the task parallelism identified in that hotspot. Later in our evaluation, it is found that this metric is not a very good indicator in the case where tasks are identified in recursive function calls. This is because DiscoPoP does not keep separate records of the recursive invocations of a function and the loss of this data effects our estimated speedup calculations.

Our output format for task parallelism identification algorithm helps programmers to easily transform code using supporting structures like master/worker or fork/join. The implementation of these supporting structures requires correct synchronization between parallel tasks. Our approach

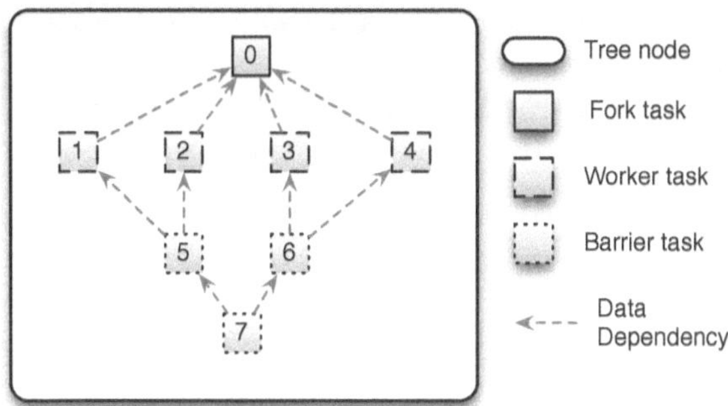



Figure 6.1: PEG of function `cilksort()` from *sort* benchmark (©2016 IEEE).

Table 6.1: Overall results for identification of the task parallelism pattern in different applications from BOTS and Polybench (©2016 IEEE).

| Application | Benchmark Suite | LOC | Exec Inst % in Hotspot | Identified Pattern |
|---|---|---|---|---|
| fib | BOTS | 32 | 100.00% | Task parallelism |
| sort | BOTS | 305 | 94.89% | Task parallelism |
| strassen | BOTS | 399 | 90.27% | Task parallelism |
| 3mm | Polybench | 166 | 99.44% | Task parallelism + doAll |
| mvt | Polybench | 114 | 91.24% | Task parallelism + doAll |
| fdtd-2d | Polybench | 142 | 76.51% | Task parallelism |

identifies these synchronization points in the form of fork, worker, and barrier CUs.

## 6.2 Evaluation

We performed the evaluation of our approach to identify task parallelism pattern on two benchmark suites, namely: the Barcelona OpenMP Task Suite (BOTS) [112] and Polybench (a collection of benchmarks with static control parts) [30]. The former suite is designed to study OpenMP-based task parallelism, while the later one contains different kernels from linear algebra and data mining. BOTS contains both sequential as well as parallel implementations of all the benchmarks in it. This allowed us to compare our identification results for applications in BOTS with their respective parallel versions. Polybench contains only sequential programs. So, we implemented all the task parallelism patterns identified by our approach, manually.

The testing environment was the same as described in the evaluation section of Chapter 5. The test hardware had 2×8-core Intel Xeon E5-2650 2 GHz processors with hyper-threading and 32 GB

Table 6.2: Comparison of estimated and achieved speedups of identified task parallelism (©2016 IEEE).

| Application | Total Instructions | Instructions on Critical Path | Estimated Speedup | Achieved Speedup | No. of Threads |
|---|---|---|---|---|---|
| fib | 52 | 16 | 3.25 | 13.25 | 32 |
| sort | 2478 | 1172 | 2.11 | 3.67 | 32 |
| strassen | 11722739 | 3349354 | 3.5 | 8.93 | 32 |
| 3mm | 3293952 | 2195968 | 1.5 | 12.93 | 16 |
| mvt | 9600 | 4896 | 1.96 | 11.39 | 32 |
| fdtd-2d | 137560 | 63309 | 2.17 | 5.19 | 8 |

memory, running Ubuntu 12.04 (64-bit server edition). Clang 3.3 was used for the compilation of benchmarks. Table 6.1 shows the overall results of the task parallelism identification.

We identified task parallelism in 6 benchmarks in total. A comparison estimated and achieved speedups of all identified task parallelism patterns is given in Table 6.2. Three of these were from BOTS and three from Polybench. Although BOTS has been developed for studying OpenMP task parallelism, we discovered that most of BOTS applications have doAll parallelism; this is implemented in the parallel version of BOTS using OpenMP tasks. We discuss the identification results for both benchmark suites in detail in the following.

### 6.2.1 BOTS

We identified task parallelism in the benchmarks *fib, sort,* and *strassen* from BOTS. All of these applications have multiple independent recursive calls that were identified as separate tasks. *fib*, as we show in Listing 6.1, calls itself twice. DiscoPoP identified both of these function calls as independent tasks. Listing 6.2 shows the simplified output of our tool with task parallelism identification. It shows that the first task starting from the start of the function is a fork task. The next two tasks are worker tasks and both are dependent on the first task. The last task is a barrier as it depends on the results of both the worker tasks.

Listing 6.1: Hotspot function of *fib* benchmark.

```
26    long long fib (int n)
27    {
28        long long x, y;
29        if (n < 2) return n;
30
31        x = fib(n - 1);
32        y = fib(n - 2);
33
34        return x + y;
35    }
```

Listing 6.2: Simplified output of the task parallelism identification for the hotspot of *fib* shown in Listing 6.1.

```
<region id="0" start="1:26" end="1:34"  workload="100.00%">
    <pattern name="task parallelism">
        <task id="0" ...>
            <Lines>1:26, 1:27, 1:28, 1:29</Lines>
            <Deps></Deps>
            <classification>Fork</classification>
        </task>
        <task id="1" ...>
            <Lines>1:31</Lines>
            <Deps>1</Deps>
            <classification>Wroker</classification>
        </task>
        <task id="2" ...>
            <Lines>1:32</Lines>
            <Deps>1</Deps>
            <classification>Worker</classification>
        </task>
        <task id="3" ...>
            <Lines>1:34</Lines>
            <Deps>1, 2</Deps>
            <classification>Barrier</classification>
        </task>
        <ES>3.25</ES>
    </pattern>
</region>
```

Listing 6.3: Parallel implementation of *fib*.

```c
long long fib (int n)
{
    long long x, y;
    if (n < 2) return n;

    #pragma omp task shared(x) firstprivate(n)
    x = fib(n - 1);
    #pragma omp task shared(y) firstprivate(n)
    y = fib(n - 2);

    #pragma omp taskwait
    return x + y;
}
```

The parallel version of *fib* has been implemented using task parallelism at the same locations.

Listing 6.3 shows the parallelized version using OpenMP task construct. We can see the one-to-one correspondence between our generated output and the parallel implementation. Our estimated speedup was 3.25; however, the parallel implementation of *fib* included in BOTS achieved a speedup of 13.25. There is such a big difference between these two because DiscoPoP does not record a function's number of recursive invocations. Our calculation is based on only one recursive step; the actual application may achieve a better speedup with the parallelism available in several recursive calls.

In Section 6.1, we discussed the PEG of *sort* in detail. Figure 6.1 on Page 104 shows the PEG of function `cilksort()` in *sort*. Here $CU_1, CU_2, CU_3,$ and $CU_4$ represent recursive calls to the function itself. $CU_5, CU_6,$ and $CU_7$ are the calls to function `cilkmerge()`.

Another task parallelism opportunity is identified in *sort* in function `cilkmerge()` with a similar PEG as in *fib*. Task parallelism patterns identified in both `cilksort()` and `cilkmerge()` have been implemented in the parallel version of *sort* in BOTS. By exploiting both task-parallelism opportunities, the parallel implementation of *sort* from BOTS achieved a maximum speedup of 3.67 on 32 threads.

In *strassen*, there are seven independent recursive calls in the `OptimizedStrassenMultiply()` function. We classified all of these as worker tasks. A loop after these seven recursive calls uses their computed results, thus becoming a barrier task. Exactly the same task parallelism pattern as the one we identified has been implemented in the parallel version of *strassen* for the same seven recursive function calls in BOTS.

### 6.2.2 Polybench

We identified task parallelism pattern three benchmarks of Polybench, namely: *3mm, mvt,* and *fdtd-2d*. In *3mm*, we identified three loops in function `kernel_3mm()` (as shown in Listing 6.4) as tasks. The first two loops were classified as independent worker tasks and the third loop as their barrier task, because it depends on the first two loops. The parallel version of *3mm* is similar to that of *fib* shown in Listing 6.3. In *mvt* two independent loops in function `kernel_mvt()` were categorized as worker tasks. All the loops that were identified as tasks in both *3mm* and *mvt* were also classified as doAll loops. We implemented combined task and doAll parallelism for both of these benchmarks and achieved speedups of 12.93 and 11.39 for *3mm* and *mvt*, respectively.

Listing 6.4: Hotspot function of *3mm* benchmark.

```
1   void kernel_3mm(){
2       for (i = 0; i < _PB_NI; i++)//worker
3           E[i] = A[i]+B[i];
4       for (i = 0; i < _PB_NI; i++)//worker
5           F[i] = C[i]+D[i];
6       for (i = 0; i < _PB_NI; i++)//barrier
7           G[i] = E[i]+F[i];}
```

We identified task parallelism in the only hotspot loop in the function `kernel_fdtd_2d()` from *fdtd-2d*. This loop consists of four CUs. The dependences among these CUs were the same as among the loops in *3mm*. The first three CUs are independent and the last CU depends on the other three. DiscoPoP classified the first three as worker CUs and the last as their barrier. After implementing the task parallelism for *fdtd-2d*, we achieved a speedup of 5.19.

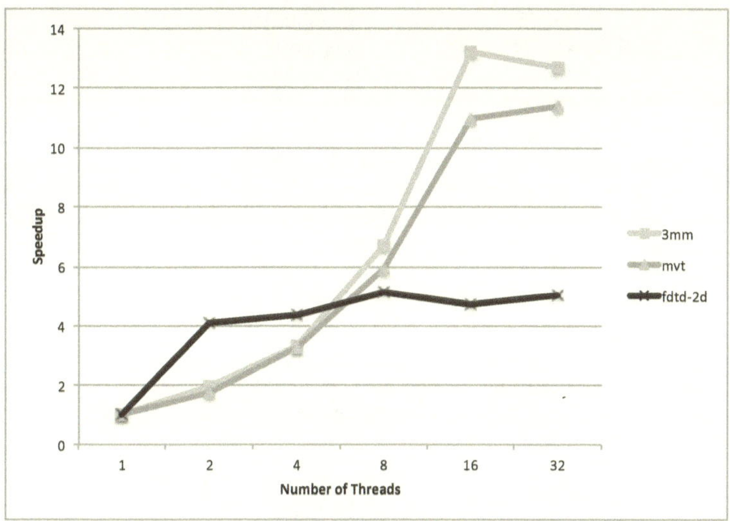

Figure 6.2: Speedups achieved for task parallelism implementation of polybench benchmarks.

We implemented the task parallelism patterns identified in its three benchmarks. The speedups for different number of threads of our parallel implementation are shown in Figure 6.2. The better speedups for *3mm* and *mvt* than that of *fdtd-2d* are due to the exploitation of doAll parallelism inside the generated tasks in these two benchmarks. This proves that our framework can identify multiple patterns simultaneously in the PEG of an application and implementing more than one identified parallel pattern at the different levels of the PEG can reduce the execution time.

## 6.3 Summary

The dynamic structure of patterns like task parallelism makes it impossible to derive a template graph or vector. The number of tasks that could be created to run in parallel at any stage of a program is highly dependent on the structure and logic of the program. Because of this, template matching cannot be used to identify such patterns.

The dependence graph of the task parallelism pattern usually has multiple tasks dependent on a single node. These dependent children tasks are usually independent from each other and thus are good candidates for running in parallel to each other. After the parallel phase, the data produced by many tasks may be consumed by a common task behaving as a barrier task.

Keeping these properties in view, we proposed using the breadth-first search (BFS) algorithm on the PEG of a program. Our modified version of the BFS algorithm classifies each CU in the PEG of a region as either a fork, a worker or a barrier task. Based on this classification, a programmer can implement the task parallelism pattern easily. We also calculate the expected speedup for each identified opportunity of the task parallelism pattern. This provides a gauge for comparison among different task parallelism identifications. However, the estimated speedup is not much accurate when the identified parallel tasks are based on the recursive function calls. This is because our base profiler, DiscoPoP, merges the data for all recursive invocations of a function into a single call. This loss of data causes our estimated speedup to be inaccurate.

The evaluation of our approach to identify task parallelism pattern demonstrates encouraging results. We analyzed different benchmarks from BOTS and Polybench. We identified task parallelism pattern in six of them. The best speedup achieved among all of these opportunities was $13.25\times$ on a maximum of 32 threads. Comparison of achieved speedups for each benchmark with the respective estimated speedups showed big differences. This is because of the recursive nature of these benchmarks. DiscoPoP does not record the number of recursive invocations, hence the estimations are calculated based on the statistics of a single recursive call.

Overall, we have shown that our approach to use modified BFS for task parallelism identification works on general purpose programs. The output of our analysis creates the basis for programmers to modify the serial programs easily into parallel ones. We did not encounter any cases of false positive or false negative pattern identifications. One of the reasons can be the small sizes of the benchmarks used for our evaluation.

# 7  Conclusion and Outlook

Software design patterns have made the life of software engineers easier, as they provide "tried and tested" solutions to recurring problems faced during different phases of software engineering. The success of software design patterns inspired the use of design patterns for parallel programming. The idea is to provide dependable and error-free solutions to different parallelization problems that are faced when developing programs that exhibit parallelism.

The application of parallel design patterns to implement parallelized software is a complex task. It requires comprehensive knowledge of the software to be developed and deep understanding of parallel design patterns. The parallelization of existing sequential applications adds another layer of complexity as the structure and working of that application is needed to be fully known beforehand. To overcome this complexity, we proposed in this book a framework that identifies appropriate parallel design patterns in the algorithm structure of a sequential application.

Our framework uses the output of DiscoPoP profiler. The profiler of DiscoPoP performs static, as well as dynamic control and dependence analysis of an sequential input application. It provides all the relevant information required to discover parallelization opportunities. The output includes data and control dependences. The profiler also divides the input program into more manageable units called computational units (CUs). All of this data is merged into a program execution graph (PEG), which in turn is used as an input to our framework that identifies parallel patterns.

Due to different algorithmic structures of different parallel patterns, we employ three separate approaches to identify appropriate patterns. The first one is based on template matching. A representative vector is derived from the structure of a parallel pattern based on its specific characteristics. Another vector is created from the PEG of a hotspot region of the input program using the same rules. The cross correlation of these two vectors indicates how effectively this pattern can be for the hotspot. We identify pipeline and doAll patterns using this approach. Our evaluation results showed successful identification of these two patterns in 12 different programs from three benchmark suites as well as an audio encoding library called libVorbis. We verified our identified patterns by comparing our results with the provided parallel versions of the programs in their respective benchmark suites, whenever possible. In our best case for pipeline identification, we implemented the pipeline pattern for libVorbis manually and achieved an average speedup of $8.4\times$ on a six core CPU machine with 12 threads.

In some cases template matching identifies the existence of a parallel pattern, but fails to provide the details, like the estimated efficiency of the parallel version if that pattern is implemented. For this reason, we employed a second approach based on linear-regression analysis. We identify multi-loop pipelines, fusion, reduction, and geometric decomposition using this approach. We record the data producing and consuming iteration numbers of either different loops (for multi-loop pipelines and fusion patterns) or a single loop (for reduction pattern) as pairs. These pairs are then used to perform linear-regression analysis to identify the patterns and their efficiency. We tested this approach on four different benchmark suites: Starbench, BOTS, Polybench, and PARSEC. The four patterns multi-loop pipeline, fusion, reduction, and geometric decomposition were identified in 11 different programs from these benchmark suites. The best speedups we achieved, on a 16 core machine with 32 threads, for the identified multi-loop pipeline and fusion patterns were $14\times$ and $16.18\times$, respectively. Also, we showed the superiority of our approach in identifying the reduction pattern over state-of-the-art static tools.

Template matching and linear-regression analysis identify the before mentioned parallel patterns. However, they cannot identify the task parallelism pattern due to the very dynamic nature of its algorithmic structure. To identify this pattern, we introduced a third approach based on the breadth-first search algorithm. The approach classifies CUs of a hotspot into fork, worker, or barrier CUs. Additionally, we provide an estimation of speedup for the identified task parallelism pattern. Our evaluation on BOTS and Polybench showed successful identification of the task parallelism pattern in six different programs. The reported estimated speedup were smaller than the actual speedup of the parallelized code due to the missing runtime information of recursive invocations in these programs by DiscoPoP. The best speedup achieved on a 16 core machine with 32 threads was $13.25\times$, where the estimated speedup of the same program was $3.25\times$.

Overall, our evaluation showed successful identification of different parallel patterns. We identified multiple parallel patterns in nested hotspots. Our approach improved the speedup by employing multiple parallel patterns when parallelizing a program. The slowdown of our approach is $62\times$ compared to the actual execution time of the program. Our approach not only suggests the identified parallel patterns but also shows how the code must be divided to fit the structure of the pattern.

Our framework could be extended to identify additional parallel patterns in the future. By extracting template vectors for new patterns, we could apply template matching to identify them. Additionally, identifying reduction pattern could be improved by inferring the reduction operators. By adding multiple loop optimization techniques like peeling and fission, the suggestions of our framework could be improved. The ultimate goal of our parallel pattern identification is to achieve automatic parallelization of sequential legacy programs. A direction of the future work could be having a tool that semi-automatically transforms a sequential application into its parallel version, with little user intervention. For this purpose, our framework can be extended to automatically map the identified patterns in the program's algorithm structure design space onto their related patterns in the supporting-structure design space. Then, this mapping can be further used for the selection of suitable implementation mechanisms.